VOL.8

JOURNAL OF ARMORED ASSAULT & HELIBORNE WARFARE

CONCORD
PUBLICATIONS COMPANY

Editor: James R. Hill

603-609 Castle Peak Road
Kong Nam Industrial Building
10/F, B1, Tsuen Wan
New Territories, Hong Kong
www.concord-publications.com

We welcome authors who can help expand our range of books. If you would like to submit material, please feel free to contact us.

We are always on the look-out for new, unpublished photos for this series. If you have photos or slides or information you feel may be useful to future volumes, please send them to us for possible future publication. Full photo credits will be given upon publication.

ISBN 962-361-074-2
printed in Hong Kong

Leopard 2A6

The Future Spearhead of the German Panzertruppe

Walter Böhm

All thirteen active Panzerbataillonen *of the* Bundeswehr *have the identical structure. The tank companies under order of a battalion can only be distinguished by their assignation to* Reaktionskräfte *or* Verstärkungskräfte*. A company is composed of three platoons, each with four KPz Leopards, and two additional tanks for the company commander and the company executive officer.*

Since its introduction in 1979, the Leopard 2 main battle tank (MBT) has been produced in eight rates and versions for the German *Bundeswehr*. In the 1970s, it was expected that the Leopard 2 weapon system would be used for the next thirty years within the *Bundeswehr* tank fleet. Today armed forces of ten European nations use the superior Leopard 2 weapon system. To bring the weapon system up to its current technical level, different improvement programs were initiated in the past, resulting in the Leopard 2A4 version.

After the end of Cold War and the dissolution of the massive threat posed by the Warsaw Pact Organization, financial restrictions caused the abortion of the development of a totally newly designed main battle tank as a replacement for the Leopard 2. Consequently, the German government decided to initiate a 3-phase improvement program for the Leopard 2. This was later reduced to a 2-step modernization program with the KWS II and KWS I measures. (KWS = *Kampfwertsteigerung* [increase in battle worthiness]).

From Leopard 2A4 to A5

After the end of the Cold War and the German reunification, the threat faced by the German *Bundeswehr* armored units changed remarkably. The army found it had to become the *Neue Heer für neue Aufgaben* (New Army for New Threats/Tasks). As a result, the concept of the *Kampfpanzer* (Kpz [battle tank]) Leopard 2A4 no longer corresponded with the tactical requirements of the *Schwere Krisenreaktionskräfte* (Heavy Crisis Reaction Forces) of the German army. Based on experiences in the Gulf War and the different conflicts in the Balkan War theater, the Germans decided to give priority to the introduction of the KWS II improvement measures and introduce the KWS I program later on. [KWS II = better armor protection and radios; KWS I = better firepower by a redesigned gun and improved ammunition.]

In the early 1990s, the KWS II improvement program created 350 Leopard 2A4s for the German *Bundeswehr* and 180 Leopard 2A4s for the Royal Netherlands Army to the Leopard 2A5 configuration. The main focus of the KWS II was the improvement of the ballistic and passive protections. By attaching modern third-generation composite armor and removable armor modules to the turret front, in addition to spall liners in the interior of the turret, the armor protection increased remarkably. Electric gun control equipment replaced the older hydraulic operating system. Also, the weapon control and night fighting capabilities of the Leopard 2 were fundamentally improved thanks to the integration of a thermal sight for the commander's periscope (PERI-R 17 A2). The rangefinder/fire control computer (FLT A2) and the turret telescopic sight (FERO-Z18A2) also were upgraded. The driver's hatch was modified and a TV camera was installed on the rear of the vehicle for safer reverse driving. With all these measures, the combat weight of the Leopard 2 increased to 59.7 tons in the A5 version.

The *Panzertruppenschule* based in Munster received their first KPz Leopard 2A5s on 30 November 1995. The first battalion equipped with the Leopard 2A5 was *Panzerbataillon* 33 in December 1995.

Leopard 2A6 KWS I
(new weapon [L55] and new ammunition [LKE II DM 53])

The next step in the evolution of the successful Leopard 2 weapon system was the KWS I improvement program. Its main focus was increasing the already enormous firepower of the Leopard 2A4/A5. The KWS I improvement program offered a redesigned gun with a longer gun barrel (from 44 calibers long to 55 calibers long) and the introduction of improved ammunition. The first batch of KPz Leopard 2A6s was ordered in the summer of 2000.

Panzerbataillon *403, which is based in Schwerin, was the first of the German* Bundeswehr *armored units to receive the advanced KPz Leopard 2A6. At the time of this writing, the Netherlands, Switzerland, Austria, Denmark, Norway, Sweden, and Spain were going to introduce the Leopard 2 MBT weapon system in different versions. Finland, Greece and Poland were also to equip their armored units with the Leopard 2 MBT.*

The KPz Leopard 2A6 is the result of the combination of KWS I and KWS II. The initial plan was to re-equip 225 main battle tanks to the A6 standard with an option for an additional second batch with 125 A6 main battle tanks. It was intended to bring all 350 KPz Leopard 2A5s to the A6 configuration. The first step was to modernize 65 vehicles (A4 version ready for modernization to the A5 level) directly from the Leopard 2A4 version to the Leopard 2A6. After this, Bundeswehr maintenance units would upgrade the Leopard 2A5 that was already in use to bring them to the A6 standard beginning in mid-2002.

The newly developed L55 gun barrel is 1.30 meters (4.3 feet) longer than its predecessor, the L44, which is used in the Leopard 2A4. The combat tactics for the Leopard 2A6 are no different, but the crew must pay attention to some peculiarities when operating the new vehicles, especially when preparing the tank for railway transport, road marches and recovery operations. A new thermal sleeve made of aluminum should minimize the thermal warp of the longer barrel. The middle and front part of the barrel have been reinforced for the use of the new LKE II DM 53 round, which creates higher gas pressure. All changes to the gun were made to prepare

In the future, the battalions of the German Panzertruppe *will be structured as follows: Four combat companies with 14 KPz Leopard 2s each, a* Stabs und Versorgungskompanie *(Headquarters and Supply Company), a* Kommandeursgruppe *(Commander's group) with two additional KPz Leopard 2s for the battalion commander and the executive officer. The entire battalion would have a strengh of 58 KPz Leopard 2s.*

The Leopard 2A6 is the result of the combination of two separate combat improvement programs, the KWS I (Kampfwertsteigerung I) and KWS II (Kampfwertsteigerung II)*. The main element of the KWS II was the improved armor protection offered from removable add-on armor on the front and sides of the turret. The result was the Leopard 2A5 configuration. The KWS I focused on increasing firepower by the integration of the new, longer L55 gun barrel.*

"Strike like Fire." With the combination of the new long L55 gun barrel and the new LKE II DM 53 ammunition, the KPz Leopard 2A6 can destroy hostile armored vehicles and tanks at long combat distances. It can also penetrate vehicles with double-reactive armor. The muzzle velocity of the rounds fired by the L55 barrel is 1,750m/s. (Carl Schulze)

it for the new LKE II ammunition, but it is still possible to fire the earlier DM 33 rounds, too. However, the enhanced penetration characteristics of the new rounds are only possible in combination with the longer L55 gun barrel.

The penetrator of the 120mm LKE II DM 53 ammunition was developed to penetrate double-reactive armor. It is made of a wolfram base (WSM IV) and is up-to-date in modern conventional anti-tank ammunition (without using depleted uranium). The effective combat range of the Leopard 2A6 was increased up to 4000 meters (4372 yd). With the introduction of the Leopard 2A6, the German *Bundeswehr* also introduced the new and improved ammunition. Compared to the first Leopard 2 tanks in 1979, the firepower and pass-through capacity of the Leopard 2A6 in 2000 had increased 100 %.

The First Leopard 2A6 in German Panzertruppe Units

In November 2001, the first unit of *Bundeswehr Heer* (German Army Land Forces) re-equipped with the modern Leopard 2A6. That unit was the 5./PzBtl 403. The 5th company has a total of 14 *Kampfpanzer* Leopard 2A6s dispersed among three platoons, each with four MBTs, and two additional MBTs for the company commander and the executive officer. Future plans foresee PzBtl.403 with a total of 58 *Kampfpanzer* Leopard 2A6s, organized in four equally structured combat companies (with 14 KPz) and two KPz for the so-called *Kommandeursgruppe* (commander's group).

"Charlie Zug der 3./PzBtl 403 weicht aus." (Charlie platoon, 3rd Company, PzBtl 403 evades). Combat tactics with the new Leopard 2A6 require no changes because of the longer gun. To minimize the greater thermal deformation of the longer L55 gun barrel, the fiberglass-covered thermal sleeve of the Leopard 2A4/A5 versions was replaced with an aluminum-covered thermal sleeve on the Leopard 2A6.

The goal of the KWS I combat improvement program was the increased firepower of the Leopard 2 based on the new 120mm/L55 smoothbore main gun, developed and built by Rheinmetall. The life of the new "Seelenrohr L55" gun barrel is up to 1500 shots. (Carl Schulze)

Submerged nearly to the top of its hull, a Leopard 2A6 slices through a shallow body of water. With proper preparation, the A6 version of the Leopard 2 can ford a body of water 4m (13ft) deep. (Carl Schulze)

Tanks belonging to 3./PzBtl 403 advance near the south gate of their barracks at Stern-Buchholz, Schwerin. The unit is ready for the road march to the training area at Lübtheen. When driving in a convoy, all the Leopards fly blue flags, while the last one in a column is identified by a green flag.

Before the Leopard 2A6 column leaves the military area to travel on civil roads, a last briefing with the civil police is necessary. Notice the new L55 gun barrel juts out over the rear of the tank. Because of this, the end of the gun barrel is fitted with a reflective cover.

Earlier plans envisioned PzBtl 403 as a mixed unit with Leopard 2A5s and Leopard 2A6s together in the line, but this plan was obsolete. In 2002, the German *Bundeswehr Heer Panzertruppe* still had four different versions of the Leopard MBT tank family in use: the older but still reliable Leopard 1A5, the Leopard 2A4, the Leopard 2A5, and the Leopard 2A6. When all improvement measures were completed, the German Army had a Leopard tank fleet with 350 Leopard 2A6s and 1134 Leopard 2A4s deployed to 13 equally structured active tank battalions. These units are under order of the *Reaktions* (crisis reaction forces) and *Verstärkungseinheiten* (reinforcement units).

Euro Leopard 2A6

The European nations using the Leopard 2 in their armies looked forward to improving and modernizing their tank fleets, following the technical progress and the new challenges on future battlefields. The Leopard 2 should be a superior weapon system in the future, as well. At the time of this writing a reliable mine protection kit was being developed and tested for use by the German, Dutch, Swiss, Swedish, and Norwegian tank units. By the end of 2002, the first German Leopard 2A6s were to be fitted with effective protection against blast and P-mines.

Tanks of 3./PzBtl 403 roll down a tree-lined street, following the leading company commander's tank. During road march operations on civil roads, the average speed of convoys is about 60km/h (37mph). The minimum distance between the tanks should be no less than 50 meters (55 yards). For security reasons, when Leopard 2A6 tanks travel on civil roads, they must always turn their longer gun barrels backward to the 6 o'clock position.

The first KPz Leopard 2A6s of 3./PzBtl 403 arrive at the training area at Lübtheen. The yellow light, which was ordered "on" for the road march to warn of the vehicle's overwide size, was removed by the crew.

Once they arrived at the training area, the crew removed the reflective cover from the gun muzzle. Now the turret is turned to the 12 o'clock position.

During the first firing of the 120mm L55 gun on the brand new KPz Leopard 2A6s, the master gunner checks the adjustment of the main gun with the ZPG (Ziellinien Prüfgerät, 10x magnification, forward).

The 120mm/L55 main gun of the Leopard 2A6 has an overall lengh of 6.60m (22 ft). There is a muzzle reference device mounted on the front end of the muzzle for in-field adjustment of the main gun.

To maintain the entire Leopard 2 weapon system at a progressive technological level, the following improvement measures have undergone tests and trials: An integrated control and information system (*Integriertes Führungs* – IFIS Information System), Identification Friend-Foe (IFF), air-conditioning system, soft-kill protection system (against guided weapons), hard-kill protection (against RPG-7 rounds, for example), and last but not least, the increasing of mobility and agility through the use of the new Euro Power Pack (engine and transmission) with its 1800hp output and lower fuel consumption.

The Leopard 2 possesses enough weight and space reserves that it is ready for even more technological challenges.

History of Panzerbataillon 403

The emblem of PzBtl 403 is divided into four sections, and features the colors blue, yellow and red. These colors were adopted from the emblem of PzGrenBrig 40 and the regional colors of the German state of Mecklenburg-Vorpommern. In the upper left part of the emblem is an iron cross, and in the upper right portion is the symbol of the Mecklenburger Stier (bull). The symbol of the bull is a sign of natural power and the attachment of the military unit to the country. The silhouette of the tank represents the kind of military unit. The beech tree in combination with the star represents the home base of PzBtl 403, the village of Stern-Buchholz.

Panzerbataillon 403 was activated on 2 April 1991. The unit is based on the structure of the MotSchRgt 27 (Hans-Kahle-Regiment), a unit in the *Nationale Volksarmee* (NVA), the army of the former East Germany, After the reunification of West and East Germany, the *Bundeswehr* adopted the former NVA unit to transfer it to the so-called *Armee der Einheit*. The equipment of MotSchRgt 27, including its main weapon systems AIFV BMP 2 and MBT T-55, was transported to storage depots to await destruction according to the KSZE contracts. Most of the former NVA officers and NCOs were adopted into the new *Panzerbataillon* 403. Like all personnel, they were trained and educated to use the improved Leopard 1A5 main battle tank. Under the KSZE memorandum, many tank battalions in West Germany were to be deactivated. The units handed over their equipment and vehicles to the newly formed units in the East German states, such as the new PzBtl 403.

Introduction of KPz Leopard 1A5 by PzBtl 403

PzBtl 403 belongs to *Panzergrenadierbrigade* 40. On 3 May 1991, the first Leopard 1A5s were delivered to PzBtl 403 and stationed at the barracks in Stern-Buchholz. In addition to the task of training the soldiers for the new Leopard 1A5 weapon system, it was a challenge to rebuild and repair the old barracks. On 16 April

A new supply of ammunition for the 120mm main gun. The 120mm KE (kinetic energy) rounds shown here must be passed carefully by one crew member to another because rough handling can damage the round. The training rounds are colored blue, while the combat ammunition is black. The arrowhead rounds, which are called APFSDS (Armor Piercing Fin Stabilized Discarding Sabot), penetrate the armor of the target and destroy the interior of the hostile tanks with an explosion of splinters.

Like all versions of the KPz Leopard 2, A6 vehicles have an ammunition supply of 42 rounds of MZ or KE ammunition. Fifteen rounds are stored on the left side of the turret, and the other 27 are stored forward of the driver's compartment. The 120mm rounds have semi-combustible cartridge cases. After the round is fired, only the metal base stub is ejected into a box under the breech.

1993, the newly educated crews had their first live fire with their Leopard 1A5 at Lübtheen training area.

In January 1996, elements of PzBtl 403 participated in a CMTC Hohenfels rotation, supporting units of PzGrenBrig 41. The CMTC Hohenfels is a U.S. installation under order of 7th U.S. Army. In March 1997, elements of PzBtl 403 were deployed to the new *Bundeswehr Gefechtsübungszentrum Heer* (GÜZ), the combined arms training center for the German army units, at Letzinger Heide near city of Magdeburg. The last live fire exercise with Leopard 1A5s was held in May 1997 at the Bergen-Hohne training range.

Re-equipment with the Leopard 2A4

At the end of March 1997, the first eight KPz Leopard 2A4s were delivered to PzBtl 403. For a short time the unit was equipped with Leopard 1A5s and Leopard 2A4s. The first live firing with Leopard 2A4s took place in April 1997 at the Bergen-Hohne range. By the end of September 1997, PzBtl 403 was completely re-equipped with the Leopard 2A4 weapon system. In July 1998, the unit participated in the "Altmärker Rochade" FTX with their Leopard 2A4 tanks. This exercise was held at the *Gefechtsübungszentrum Heer* (GÜZ) near Magdeburg.

In September/October 1998, elements of PzBtl 403 were involved in an exercise at CMTC Hohenfels. In the second half of 1999, the 5th company of PzBtl 403 was deployed to Kosovo for a KFOR mission. The unit was a part of the 2./vst Mech Battalion. The soldiers and crewmen were trained before this mission to familiarize them with the new Leopard 2A5 MBT since it was the standard main battle tank in the *Bundeswehr* KFOR units. *Bundeswehr* KFOR units that were deployed near the town of Prizren and were tasked with a wide range of missions, including guarding crossroads and building up checkpoints.

Re-equipment with the Leopard 2A6

Based on the new structure of the *Bundeswehr Heer*, today's PzGrenBrig 40 will be deactivated in the near future. Then PzBtl 403 will be under order of the PzGrenBrig 41, as a part of the 14th PzGren Division. On 6 March 2001, the first Leopard 2A6 was handed over by Krauss-Maffei in a special ceremony in München. The PzBtl 403 is the first unit in *Bundeswehr Heer* to be equipped with Germany's most modern main battle tank.

The 5./PzBtl 403 was deployed to Münsingen training area by the end of 2001. There the tank first fired training ammunition. At the same time the battalion received more Leopard 2A6s and left behind the old Leopard 2A4. In June 2002, the 3./PzBtl 403 deployed to the Bergen training area for the first firing with the new tank. The other companies of PzBtl 403 were scheduled to re-equip from the Leopard 2A4 to the new A6 vehicles in the summer of 2002.

The penetrator of the new LKE II DM 53 ammunition for the 120mm/L55 main gun of the Leopard 2A6 is developed from a wolfram alloy. These rounds do not need to use depleted uranium 235DU. During the Gulf War, the U.S. Army used rounds with the depleted uranium penetrator. The so-called Silver Bullet rounds for the M1A1 MBT were very effective against Iraqi T-72s. Similar to this, the wolfram penetrator of the German DM 53 rounds passes through armor and sets fire to the inside of the hostile vehicle.

KPz Leopard 2A6s of Charlie Zug, 3./PzBtl 403 were uparmed with MZ practice rounds (MZ = Mehrzweck (multi-purpose)). The MZ, or HEAT (High Explosive Anti-Tank), round is a hollow-head type of ammunition. These rounds are used against soft and hard targets. In WWII, the German tanks fought very successfully with hollow-head ammunition against Russian and Allied fighting vehicles. Today the hollow-head rounds are used mainly against soft or light armored targets because of the explosive and splinter effect.

Technical Data for the Leopard 2A6

Crew : 4

Dimensions:

Length with gun: 11.27m
Width overall: 3.74m
Height to top of commander's periscope: 3m
Ground clearance: 0.5m

Weight:

Combat weight: 60.1 tons
Power to weight: 25hp/t
Ground pressure: 0.93kg/cm2

Performance:

Max speed: 72km/h
Fording with preparation: 4m
Trench: 3m
Vertical obstacle: 1.10m
Range (max.): 550km
Range (cross country): 240km

Engine: 12-cylinder, turbo-charged, diesel MTU-MB 873 with two turbo-chargers with cubic capacity of 47.6 liters; output of 1100kW/1500hp

Transmission: 4 forward gears / 2 reverse gears

Fuel storage: 1,200 liters

Fuel consumption: 218 liters/100km

Armament:

1x 120mm L55 smoothbore main gun
Ammunition stock of 42 rounds MZ/KE ammunition
1x coaxial 7.62mm MG3 machine gun
1x 7.62mm MG3 anti-aircraft machine gun
4750 7.62mm rounds
2 banks of eight 76mm caliber Wegmann smoke dischargers on each side of turret NBC protection system

Fire control systems:

Main gun fully stabilized (electrically operated) GPS

Sights:

Driver: Night sight equipment
Gunner: Thermal sight and laser range finder
Commander: Thermal sight in his periscope

Manufacturer: Krauss-Maffei-Wegmann, Germany

Platoon markings. The 3./PzBtl 403 uses the letters "A", "B" and "C" on the rear of the turret to identify the different platoons. Additionally, the crews fly small pennants on the radio antenna. The first platoon (Alpha-Zug) has yellow and white pennants, the second platoon (Bravo-Zug) yellow and red pennants, and the third platoon (Charlie-Zug) yellow and yellow pennants.

A FlaMG3 being uparmed with 7.62mm rounds for night gunnery. For night firing, normal rounds are packed together with tracer rounds in the ammo belts to help the machine gunner achieve better control of the effect of the fire.

This tank belongs to the company commander of 3./PzBtl 403. It is marked with a yellow-black pennant on the antenna, tactical stencilling on the left rear of the turret, and the company commander's name on the right rear of the turret.

The KPz Leopard 2A6s of Bravo-Zug, 3./PzBtl 403 fly a yellow and red pennant. The tank's longer gun barrel forces the crews to pay special attention when driving in woods or on narrow roads. Like the Leopard 1, the Leopard 2s of the first production lots were fitted with an ammunition resupply hatch on the left side of the turret. Later on, after the introduction of the A2 production batch, this weak point in the turret's ballistic protection was eliminated by welding this hatch closed.

When the Leopard 2s are driven in reverse, the mirrors of the muzzle's field adjustment device often get dirty. At the time of this writing, the German tank units were testing a Swedish solution consisting of rubber skirts mounted on the front of the hull for keeping off dust and dirt. Since the eighth production batch, the Leopard 2 has been fitted with a field adjustment device on the muzzle of the main gun. All earlier production vehicles will be modernized with this equipment.

Members of the Leopard 2 family of vehicles. The AVR Bergepanzer Büffel has the same agility and mobility as the Leopard 2 MBT. More then 60 percent of the Büffel's parts and components were adopted from the Leopard 2. The South Korean K1 recovery tank and the French AVR based on the Leclerc MBT both use components from the German Büffel AVR.

Due to the longer gun barrel of the Leopard 2A6, the procedure for recovering broken down Leopard 2A6 MBTs had to be modified. In contrast to the recovery and hauling procedure for the older Leopard 2A4/A5 variants, the A6 must be hauled from the rear, mounting the pull bars on the hull's rear eyelets. The turret must be traversed to 6 o'clock position and the main gun must fixed at maximum elevation.

A Bergepanzer 3 Büffel ARV lifts the powerpack from a Leopard 2A6. A well-trained Büffel crew can remove the complete powerpack within 30 minutes.

The different fuel tanks of the KPz Leopard 2 have a total capacity of 1,200 liters (317 gallons). Here a new Strassentankwagen 8x8 Iveco Typ MP 340 E 42 W of Transportbataillon 133 fuels a tank platoon. Inspired by experiences of the Bundeswehr's deployment in the Balkan theater, the STW 8x8 was introduced four years ago. The fuel tank has a capacity of 18,000 liters (4,752 gallons). Within a minute, more than 180 liters (47.5 gallons) of fuel can be pumped to other vehicles.

The design of the Leopard 2A6 lends itself well to camouflage. From the ground to the commander's periscope, the tank is only 3 meters (9.8 feet) high. (Ralph Zwilling)

Together with the German Bundeswehr, the Royal Netherlands Army will upgrade 180 Leopard 2 main battle tanks to the A6 standard. The improvement measures performed on the Leopard 2A6s of the Dutch forces are identical to those of the program for the Bundeswehr's Leopard 2A6s. The Royal Netherlands Army will also obtain the new LKE II DM 53 ammunition for the new gun. (Ralph Zwilling)

This photograph shows the typical armament of the Leopard 2A6 main battle tank. Visible are the long 120mm L55 smoothbore main gun, the 7.62 machine gun and the banks of eight 76mm Wegmann smoke dischargers on the left side of the hull. Another set are located on the right side of the hull, and it appears that more smoke dischargers are also affixed to the front of the hull. (Ralph Zwilling)

Along with a generous amount of camouflage material, this Leopard 2A6 features a leather skirt at the front of the hull. This was attached in an attempt to prevent dust and dirt from dirtying the muzzle field adjustment device while the tank is being driven in reverse. (Ralph Zwilling)

Rear view of a Leopard 2A6. The baskets on the rear of the turret and the TV camera that provides for safer reverse driving were modification elements of the A5 version of the Leopard 2. (Ralph Zwilling)

The hull of all the Leopard variants consists of three sections: the driving compartment in the front, the fighting section in the middle and the engine located in the rear in the rear section of the MBT. The transmission in the Leopard 2A6 consists of four forward gears and two reverse gears. The maximum range that the tank can achieve is 550 km (342 miles). (Ralph Zwilling)

In an effort to obscure its position, a Leopard 2A6 expels a cloud of smoke from one of the many smoke dischargers located around its turret. (Ralph Zwilling)

The commander's hatch is located on the right side of the turret. In front of the hatch may be seen the gunners's stabilized EAMES-15 sight. The PERI-R 17 A2 panoramic sight is positioned to the rear left of the commander's hatch at the far right in the photo. This useful equipment assists in day/night observation and target acquisition. It also provides a 360 view of the terrain. Note the tanker's helmet and headset. (Ralph Zwilling)

Two crew members keep a vigil in the commander's and loader's hatches during an exercise. The weapon is a FlaMg 7.62mm MG, for which there are 4800 rounds available. (Ralph Zwilling)

A tank commander directs the operation of a Leopard 2A6. The insignia worn on his beret is noteworthy. (Ralph Zwilling)

This photograph of a crewmember of Leopard 2A6 provides a good study of the goggles and communication gear worn by the tankers. The Leopard 2A6 has four crewmen: commander, gunner, loader, and driver. (Ralph Zwilling)

The Leopard 2A6 is capable of displaying impressive performance. It can achieve a maximum speed of 72km/h (45mph) and overcome obstacles up to 1.1m (3.5ft) high. (Ralph Zwilling)

Emblem of PzBtl 403. The emblem, which is divided into four sections, features the colors blue, yellow, white, and red. In the white quarter is an iron cross, and in the yellow portion is a bull. The symbol of the bull is a sign of natural power and the attachment of the military unit to the country. The representation of a tank in the pinkish-red quarter represents the kind of military unit. The beech tree and the star in the blue section represents the village of Stern-Buchholz, the home base of PzBtl 403. (Ralph Zwilling)

When traveling cross-country, the Leopard 2A6 has a range of 240km (150 miles). Note the modern Iron Cross national insignia on the side of the turret. (Ralph Zwilling)

The emblem of 4./PzBtl 403 consists of a shield of sky blue and white with a gold griffin at the top, the silhouette of a tank in the middle, and a black rectangle below. Note also the red and black pennant of the company. (Ralph Zwilling)

One of the finest main battle tanks in the world, and one that many European nations desire for their own armored forces, the Leopard 2A6 features the longer L55 gun, an auxiliary engine, improved mine protection, and an air-conditioning system. (Ralph Zwilling)

DAGGER BATTALION

The 54th Engineer Battalion (C) (M)

Ralph Zwilling

Because the dimensions of the M60A1 AVLB are too large for rail tunnels, the launcher is lowered onto the rail car and the bridge has to be transported separately by trucks. The M60A1 AVLB is fixed onto the rail car only with some chock blocks. The M60A1 AVLB improvement program consists of three lots (1st: 23 tanks, 2nd: 20 tanks, 3rd: 25 tanks) and an optional 4th lot of 36 vehicles. The U.S. Army is currently considering a reduction of the number of improved vehicles from 68 to 56 because of funding.

Today's engineers are more than soldiers who lay and remove mines or build obstacles made of concertina wire to stop enemy ground forces on the battlefield. Though they still use the well-known Bangalore torpedo as a wire-breaching device, they are also equipped with remote controlled mine-detecting and -clearing vehicles, armored vehicle launched bridges, dump trucks, armored combat earthmovers, and rocket-propelled explosive line charges to name just a few. Among others tasks, they conduct mobility, countermobility, survivability, and general engineering support from the forward line of own troops through the communication zones to the airports or seaports of entry.

Engineers do not fight alone. They are an important part of the combined arms team, and their battlefield functions increase the team's combat effectiveness. The structure of the engineer force has been

The M9 ACE, which was introduced into the U.S. Army inventory in late 1989, is a highly mobile, armored combat earthmover capable of supporting forces in both offensive and defensive operations. It is designed to operate in forward areas of the battlefield as a general purpose engineer vehicle.

The M9 ACE is equipped with a unique hydropneumatic suspension system that allows the front of the vehicle to be raised, lowered, or tilted to permit dozing, excavating, rough grading, and ditching functions. The principal components are eight high-pressure hydraulic rotary actuators that are connected to the road wheel stations. During high-speed travel, this system assures a smooth ride through the use of shock-absorbing accumulators.

The entire 54th Engineer Battalion was deployed to Camp Virginia in the Central Command (CENTCOM) area of responsibility (AOR) at the end of January 2003. This M9 ACE tows a M105A2 $1\frac{1}{2}$-ton cargo trailer through the Bamberg local training area toward the railhead during deployment operations. Note the folded steel dozer blade.

developed to support various missions from front to rear, with more survivable and mobile forces in the forward lines. Any engineer unit could find itself pushed forward if the mission dictates. A wide variety of engineer units provide the specific technical capabilities that are required to accomplish essential, diversified tasks throughout the depth of the theater.

The corps is the U.S. Army unit capable of operating at both the tactical and operational level of war. It is the instrument by which higher echelons of command conduct maneuvers at the operational level. A tailored engineer brigade is commonly available to the corps to add more weight to its main effort and to perform special combat support functions. The 130th Engineer Brigade, based in Hanau, Germany, belongs to the

A M9 ACE drives onto a rail car during deployment operations in January 2003 in Bamberg, Germany. It only took about 4 hours to load an entire engineer line company on the German Bundesbahn rail cars. The dozer blade can be folded onto the apron for more ground clearance during travel. The chock blocks attached to the apron are used to secure the vehicle to the rail car.

A M9 ACE performs a digging operation at night in the CMTC-Hohenfels in October 2002. (CPT John Hudson)

The Small Emplacement Excavator (SEE) is a lightweight, all-wheel-drive, diesel-engine-driven, high mobility vehicle. It is fitted with a backhoe, a bucket loader, and other attachments such as a handheld hydraulic rock drill, a chain saw, and a pavement breaker. The air transportable vehicle can travel at speeds up to 80 km/h (50 mph) on improved roads, but it has limited off-road mobility. The SEE is used to rapidly dig combat emplacements such as crew-served weapon positions, command posts, and individual fighting positions. The high mobility of the SEE makes it an earthmoving machine capable of rapid movement between battle positions. Note the Falling Objects Protective Structure (FOPS) over the cab and the tactical marking on the door. The front loader has a lift capacity of 1,497 kg (3300 lb).

United States Army Europe's V "Victory" Corps. This brigade commands and controls all engineer support to the corps and is assigned all engineer units that are not organic to divisions and separate brigades.

The 130th Engineer Brigade provides mobility, countermobility, survivability, and general engineering support to the corps. It rapidly deploys to support V Corps as a Contingency Force in support of European Command (EUCOM) and Central Command (CENTCOM) regional military objectives or in support of NATO and UN military operations. It also provides trained and ready forces in support of Commander in Chief U.S. Army Europe and V Corps validated operational requirements. The 130th Engineer Brigade currently consists of the Headquarters and Headquarters Company (HHC) in Hanau, the 94th Engineer Battalion (Heavy) in Vilseck, the 565th Engineer Battalion (Provisional) in Hanau, and the 54th Engineer Battalion (Combat) (Mechanized) in Bamberg, which we will look at more closely in the following report.

On order, the 54th Engineer Battalion "Dagger Battalion" deploys all or part of its strength to conduct mobility, countermobility, survivability, and general engineering support to V Corps units. In addition, on short notice it can deploy elements in the EUCOM and CENTCOM area of responsibility (AOR) in support of V Corps to provide trained, ready forces to meet operational requirements. The 54th Engineer Battalion (Combat)

The SEE is basically a backhoe and a small front bucket attached to a Mercedes Benz Unimog truck. The excavator can aid troops to create trenches, dig trash ditches, or prepare ground for underground pipelines. It is also equipped with a hydraulic chainsaw and a jackhammer. The vehicle has a ground clearance of 43cm (17 in) and a fording depth of 76cm (30 in).

(Mechanized) has a long and distinguished record of performance, not only as an outstanding engineering unit but also as a formidable fighting force.

History of the 54th Engineer Battalion (C) (M)

World War I

Since its inception in 1917, the Battalion has remained true to its motto, "Let us try, let us do." The history of the 54th Engineer Battalion began on 7 December 1917, when it was constituted as the 42nd Engineer Battalion (Auxiliary Forestry), at Camp American University, Washington D.C. General Order #108 authorized the reorganization of the 42nd into the 20th Engineer Regiment on 18 October 1918.

From the onset, the 20th Engineers was a unique organization. An assignment to what was, at that time, the largest regiment of any kind in the world was no easy task. By 18 October 1918, the unit consisted of a Regimental Headquarters, 14 Battalion Headquarters, 49 Forestry Companies, 28 Engineer Service Companies, and two attached Engineer Service Battalions, totalling 268 officers and 19,385 men!

Following its organization, in 1918 the 20th Engineer Regiment deployed to Noisier, France, as a part of the American Expeditionary Forces (AEF). For the next two years the soldiers of the 20th toiled

A sapper operates a Small Emplacement Excavator with the help of night vision goggles. (CPT John Hudson)

This 27.5-ton M985 HEMTT (Heavy Expanded Mobility Tactical Truck) belongs to the Alpha Company's assault and obstacle platoon. The Detroit Diesel 8V92TA engine with 8 cylinders, 12-liter displacement and turbocharger provides 450 gross horsepower. Note the air cleaner situated behind the cab. The Allison HT740/4 automatic transmission has one reverse speed and four forward speeds. The difference between the M985 and the M977 is the improved Grove MHC 985 material-handling crane with its maximum capacity of 2,454kg (5,411 lb).

endlessly, constructing roads, barracks, airfields, placing mines and obstacles, and installing bridges in support of the AEF.

World War II

After the end of WWI, the 20th Engineers was deactivated at Camp Merritt, New Jersey, on 7 July 1919. The 42nd Engineer (General Service) (Regiment) was reconstituted on 1 October 1933 as an inactive unit of the Regular Army. On 1 June 1940, it was activated with this same designation at Fort Benning, Georgia, and one month later was reorganized and redesignated as the 2nd Battalion, 20th Engineer Regiment. After completing basic training, and two major construction projects in Louisiana and Mississippi, the Regiment deployed overseas aboard the *Hugh L. Scott* from Norfolk, Virginia on 22 October to assist in halting Hitler's march across Europe. Assigned as part of the Western Task Force, whose mission was to make assault landings on the coast of Africa, the 20th Engineers embarked on a course that would earn the unit its place in the annals of American military history.

In November 1942, the 2nd Battalion of the 20th Engineer Regiment landed at the port of Fedala, French Morocco, as part of the 3rd Infantry Division "Rock of the Marne", successfully assisting the infantry in securing and clearing the beachhead. In mid-November of the same year, select elements of the regiment participated in the assault landings at Casablanca. In March 1943, the Regiment was assigned as corps engineers in support of II Corps. During this time, the 20th constructed a regimental rifle range, performed extensive mine clearing operations, and constructed a major road network in support of the British 8th Army, extending its supply lines by some 20 miles. On 10 July 1943, following extensive assault training with the 3rd Infantry Division, the 20th participated in the invasion of Sicily. The unit was then assigned as 7th U.S. Army Engineers and tasked with constructing numerous timber and Bailey bridges, plus several major runways. In November of 1943, the Regiment departed

The Cummins NHC 250 6-cylinder diesel engine with 250 horsepower provides the 9.8-ton M923 with a maximum speed of 80km/h (50 mph). The automatic transmission has five forward and one reverse gear. This "Outlaw" M923 dump truck tows a M149A1 Water Trailer that has a capacity of 400 gallons.

Each engineer line company contains four 11.75-ton M929 dump trucks, which are organized in the combat engineer platoons and the assault & obstacle platoon. The M929 dump truck can carry 4.5 tons material cross-country. Note the tailgate with the tailgate wings and the yellow bridge, which is a tactical marking used only by the 54th Engineer Battalion.

The M944A1 is used as a maintenance truck in the headquarters and headquarters company of the 54th Engineer Battalion. Among many other tools, there is a lathe installed in the cargo compartment behind the cab. The soldiers call the M944A1 the "Batmobile", as you can fold the sides of the rear compartment to the top.

Tunisia, with 2nd and 1st Battalions stopping in Palermo and Trapani, Italy, where they constructed two airstrips before rejoining the remainder of the Regiment in England on 23 November.

On 15 January 1944, while on assignment in England, the 20th Combat Engineer Regiment was re-formed into the 1171st Engineer Combat Group. The old 1st Battalion, 20th Engineer Regiment, became the 20th Engineer Combat Battalion. The former 2nd Battalion, 20th Regiment, became the 1340th Engineer Combat Battalion. Both the 20th Engineer Battalion and the 1340th Engineer Battalion were assigned to the 1171st Engineer Group. It is through the 1340th Engineer Combat Battalion that the 54th Engineer Battalion traces its lineage.

On 6 June 1944, two elements of the 1171st Group (the 20th Engineer Battalion and the 1340th Engineer Battalion) participated in the D-day invasion of Normandy. In the following months, the group fought in France and Germany with tremendous success. When the formal surrender of Germany became effective on 7 May 1945, the majority of the 1171st Engineer Group was in Czechoslovakia supporting the 1st Infantry Division and the 16th Armored Division. The 20th Engineer Battalion remained on occupation duty with the 1st Infantry Division in Czechoslovakia until the 2 January 1946, when the 1340th was returned to the states and deactivated at Camp Kilmer, New Jersey.

The M48A5 armored vehicle launched bridge (AVLB), with a length of 11.28 meters (37 ft), a width of 4.0 meters (13 ft) and a height of 3.84 meters (12.6 ft), was designed to launch and retrieve a class 60 scissors-type bridge. It consists of three major sections: the launcher, the main body and the bridge.

The launcher is mounted as an integral part of the main body, which is based either on the M48A5 or the M60A1 tank assembly. It is powered by an air-cooled, turbocharged 750 hp 12-cylinder Teledyne Continental AVDS-1790-2D diesel engine that provides a maximum speed of 48 km/h (30 mph).

1945 - 1990

On 15 September of 1948, the 1340th Engineer Battalion (redesignated the 8th Engineer Combat Battalion) was reactivated at Panzer Kaserne, Böblingen, Germany. On 1 December of the same year, the unit was redesignated the 54th Engineer Battalion and assigned to support three armored cavalry regiments in the Constabulary until late 1950. In October 1953, the Battalion moved to Fliegerhorst Kaserne, Leipheim, Germany, and on 5 June 1953, it was designated the 54th Engineer Battalion (Combat). The Battalion departed Germany on 15 April 1957 aboard the *USS Buckner* as part of Operation Gyroscope. It was subsequently assigned to the 2nd U.S. Army and attached to the U.S. Armor Center at Fort Knox, Kentucky. In March of 1958, the 54th was designated a Strategic Army Corps (STRAC) unit and attached to the XVIII Airborne Corps, the STRAC Headquarters.

The 54th once again deployed to Europe during the Berlin crisis as a part of Operation Round Out. It departed the United States on 10 October 1961, aboard the *USNS Gordon*. On 21 October 1961, it arrived at Wildflecken, Germany, which would be its home for the next 23 years. From the Battalion's arrival in 1961 until the opening of the East-West border in November 1989, the Battalion had the mission as the V Corps Covering Force Engineers in support of the 14th, and then the 11th, Armored Cavalry Regiments. In 1986, the 54th Engineer Battalion became the first corps combat engineer battalion to become mechanized.

1990 - 2003

On 24 December 1990, the 54th was notified to prepare to deploy to Saudi Arabia in support of Operation Desert Shield as part of VII Corps. The 54th was attached to the 1st Armored Division "Old Ironsides" on 1 February 1991. The ground offensive commenced on 24 February, and during the 87 hours of the division attack, over 244 kilometers (152 miles) were covered, hundreds of enemy vehicles were destroyed and 2100 EPWs were captured.

Following the cease-fire, the Battalion was moved to northern Kuwait, and then later back into Iraq, as it conducted extensive denial missions destroying enemy equipment. The Battalion redeployed to Germany from 28 April through 5 May, and its equipment returned to Germany in late July through September. The 54th Engineer Battalion returned to its mission of supporting primarily the 11th Armored Cavalry Regiment, along with other units in V Corps as required, as part of the smaller, more mobile U.S. Army in Europe. Delta Company, upon its return from Southwest Asia, received the ACE Mobile Force (Land) mission and immediately began training to meet the requirements of the NATO rapid deployment force. The AMF (L) company deployed to and conducted training in Italy, Turkey, England, and Denmark, and participated in rotations to CMTC with the Battalion, when available.

The 1,419-liter (375-gallon) fuel tanks provide the M60A1 AVLB with a cruising range of nearly 470 km (292 miles). The Detroit Allison Diesel CD-850-6A powershift crossdrive transmission has two forward speeds and one reverse speed. With a ground clearance of 35.6 cm (14 in), the M60A1 AVLB can climb obstacles up to a height of 46 cm (18 in) and is able to cross trenches up to a width of 214 cm (84 in).

The aluminum alloy girder bridge that folds in half is mounted on top of the main body. Through a special hydraulic system operated by the driver, the bridge is raised to a vertical position and then extended forward, opening as it proceeds forward. Bridge emplacement can be accomplished in 2 to 5 minutes, and retrieval can be accomplished in 10 minutes under armor protection.

After spending much of 1992 in the field, in 1993 the Battalion began preparing for the Engineer Restructure Initiative (ERI) transition that was to begin in February 1993. Reconfiguring under the ERI meant the Battalion was to lose one line company, one line platoon per company, all of the vehicles and equipment for each of those units, and over one third of the Battalion's soldiers. On 15 June 1993, the 54th received official notification that it was to inactivate, with an end-date in January 1994. On 15 January 1994, the Dagger Battalions held an Inactivation Ceremony at Hanau, Germany, hosted by the 130th Engineer Brigade. On 16 February 1997, the 54th Engineer Battalion was once again reactivated and began a new chapter as one of the finest engineer battalions in the Army. Today, the 54th Engineer Battalion is based in the Warner Barracks in Bamberg, which is located in northern Bavaria, Germany.

Operation Enduring Freedom 2003

On 2 January 2003, Charlie Company received the order to deploy to Camp Virginia, Kuwait, in support of Operation Enduring Freedom in the Central Command (CENTCOM) area of responsibility (AOR). Only seven days after the order, Charlie Company loaded their vehicles and equipment on trains that transported the material to Antwerp, Belgium, where it was loaded on ships. The rest of the Battalion received the deployment order on 12 January 2003 and deployed the equipment from 15 to 18 January.

Until the soldiers deployed via airplane to Kuwait at the end of January and the beginning of February, they conducted further NBC training and live firing tables in Bamberg, Germany, to be prepared for the upcoming mission against Iraq. When Operation Iraqi Freedom started on 20 March 2003, Charlie Company was attached to the 3rd Squadron, 7th U.S. Cavalry Regiment, the reconnaissance and security force of the 3rd U.S. Infantry Division "Rock of the Marne".

Today's Organization of the 54th Engineer Battalion (C) (M)

The organization of a corps engineer battalion is almost identical in capability to the division engineer battalions in an armored or mechanized division, except it has a larger staff. It is designed to conduct engineer operations in close combat and can fight as mechanized infantry when properly trained and augmented. Mechanized corps engineers provide mobility support to reconnaissance, intelligence, surveillance, and target acquisition (RISTA) forces. This support can involve breaching natural and man-made obstacles and improving trafficability of routes for cavalry regiments, field artillery, and logistics units.

During deliberate breaches at division or brigade level, mechanized corps engineers may provide engineer support to the breach force, preserving the division engineers for follow-on operations; follow and widen breaches conducted by division engineer units; or breach obstacles bypassed by division engineer units. To weight the offensive main effort, mechanized corps engineer battalions can be task-organized to division engineers. In deliberate defense, mechanized corps engineers augment division engineers in survivability and countermobility operations. The 54th Engineer Battalion (Combat) (Mechanized) currently consists of a headquarters and headquarters company ("Headhunters") and three line companies ("Outlaws", "Bulldogs" and "Charlie Rock") **(Graphic 1).**

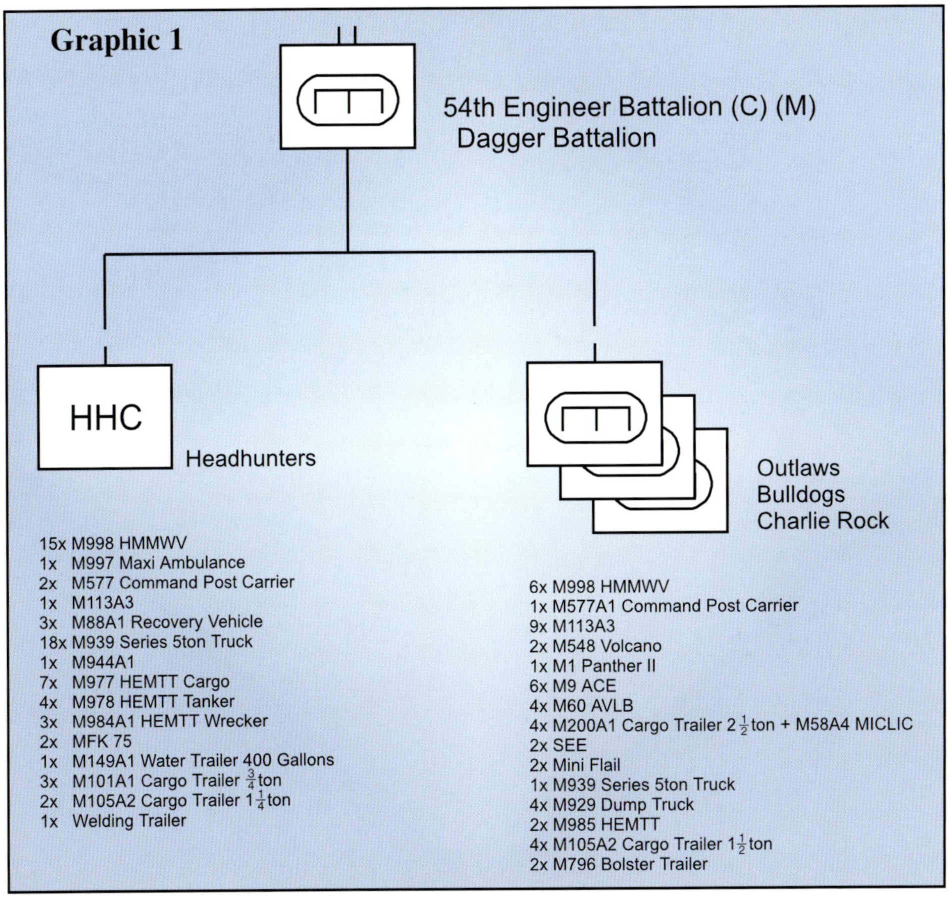

The M60A1 bridge can be retrieved from either end depending on the mission. The AVLB spans a 15-meter (49-foot) gap for Military Load Class (MLC) 70 and an 18-meter (59-foot) gap for MLC 60. The total weight of the M60A1 AVLB without bridge is 41.7 tons and 55.3 tons with bridge.

A M60A1 AVLB lays a bridge at night. The photo was taken through an AN/PVS-14 night vision goggle. (CPT John Hudson)

The headquarters and headquarters company is divided into a battalion command section and the headquarters company. The "Dagger Battalion" is commanded by Lieutenant Colonel Donald E. Jackson, who forms the battalion command section along with the Executive Officer Major Jeffrey Harrison and the Battalion Command Sergeant Major CSM Larry D. Jackson.

The company headquarters is supervised by the HHC commander, normally a captain, and his first sergeant. It provides command and control, administrative, and supply, and it co-ordinates security for the company. It is therefore equipped with two HMMWVs, two M939 series 5-ton trucks and two trailers. The administrative and logistic platoon includes the unit ministry team, the S1, the S4, and the medical section. The S1 officer and his section provide administrative and legal support and help the battalion commander provide for the welfare of the battalion's personnel. The S-4 officer supervises battalion supply. His section

These stripped hulls from two M60A1 AVLBs were photographed during the refurbishment program in the facilities of the German company Flensburger Fahrzeugbau GmbH. The hulls are entirely sandblasted before the refurbishment and they receive a new Chemical Agent Resistant Coating (CARC) afterwards. (FFG)

Most of the M48A5 and M60A1 AVLBs received IFF (Identification Friend Foe) panels prior to the deployment to Operation Iraqi Freedom. They are attached to the front and rear of the vehicles. The crew of the M48A5 AVLB consists of the tank commander and the driver located in the middle of the hull. Two M250 smoke grenade launchers mounted beside the bridge launcher at the front provide the vehicle with self-screening capability.

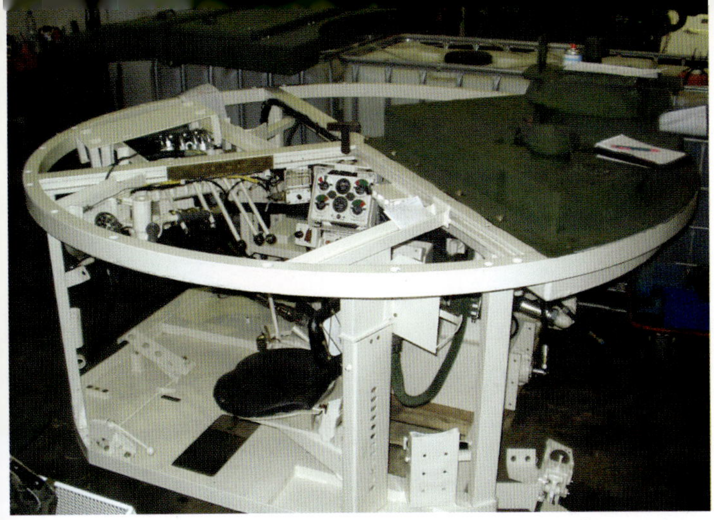

The crew compartment is seen in close-up from the vantage point of standing outside of the hull. Note the steering control, the indicator panel and the driver's seat. (FFG)

A close-up of the air-cooled, turbocharged 750 hp 12-cylinder Teledyne Continental AVDS-1790-2D diesel engine. (FFG)

coordinates the overall supply activities of the battalion and works closely with the operations section.

The chaplain's assistant, who drives a HMMWV, is the only member of the unit ministry team. His mission is to assist the unit chaplain in providing religious services and soldier welfare ministries. Battalion medical support is provided by the medical section, which is only equipped with a M997 Maxi Ambulance HMMWV.

The communications section (S6), with its two HMMWVs and a cargo trailer, maintains and repairs the innumerable battalion computer and communication systems.

The food service section provides food service support to all battalion elements with its two M939 series 5-ton trucks and two MKT-75 Mobile Field Kitchen Trailers.

The "Dagger Battalion" support platoon has two HMMWVs, seven M977 HEMTT Cargo and four M978 HEMTT Tanker.

The operations section (S-3) and intelligence section (S-2) form the tactical operations center (TOC), which coordinates the Battalion's operations. Its vehicle park includes two M577A2 command post carriers (CPC), two HMMWVs and two cargo trailers.

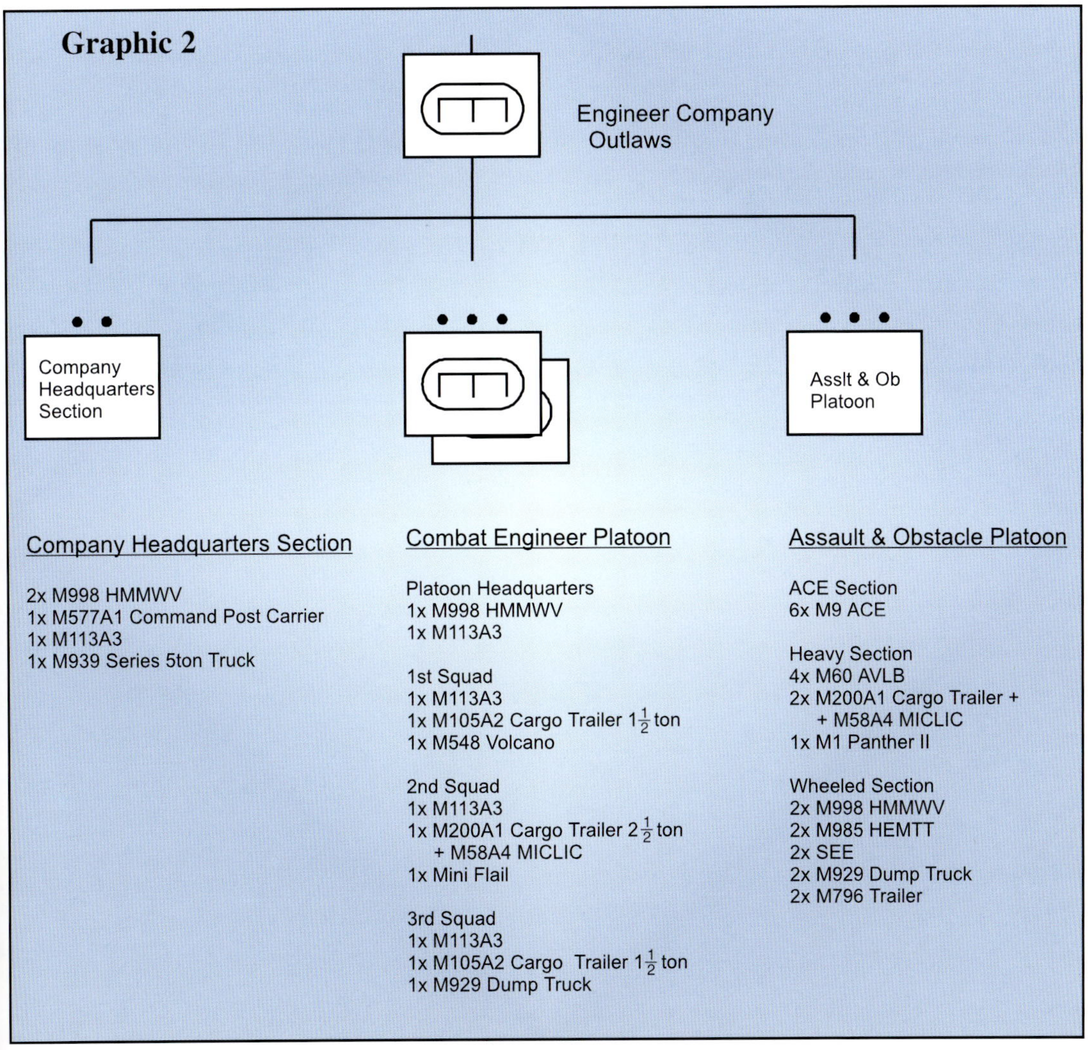

The maintenance platoon, with its two HMMWV, three M88A1 Armored Recovery Vehicles, three M984A1 HEMTT Wrecker, twelve M939 series 5-ton trucks, a M944A1 "Batmobile", and a welding trailer performs unit maintenance for all organic battalion elements. The responsibility for operator and crew maintenance remains with the companies **(Graphic 2).**

The three engineer line companies, which are organized identically, consist of a company headquarters, two combat engineer platoons and an assault & obstacle platoon. The company headquarters, which provides command and control of the unit, as well as supply and NBC support for unit combat engineer operations, is equipped with two HMMWVs, one M939 series 5-ton truck, one M113A3 Armored Personnel Carrier (APC), and one M577A1 Command Post Carrier. Each combat engineer platoon has a platoon headquarters with a HMMWV and a M113A3, and three engineer squads. The 1st Squad consists of a M113A3 with a M105A2 cargo trailer and a M548 Volcano Mine Dispenser. The 2nd Squad has a M113A3 with a M200A1 $2\frac{1}{2}$-ton cargo trailer and

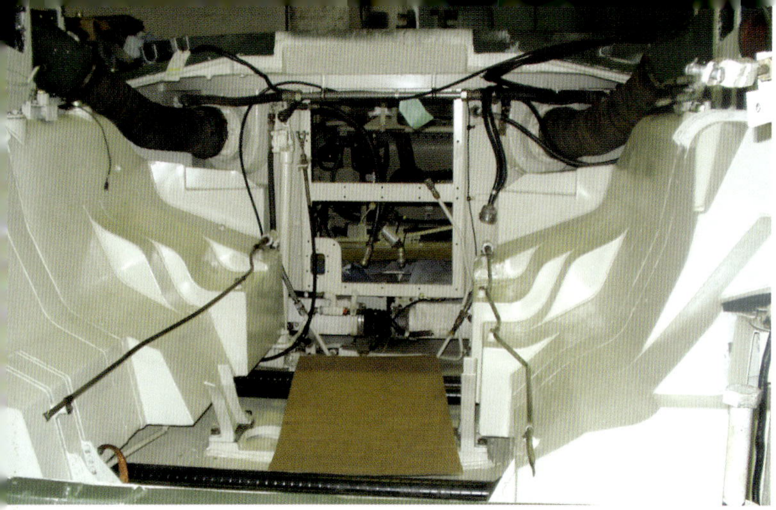

A look inside the engine compartment of a refurbished M60A1 AVLB. (FFG)

Some Heavy Equipment Transport Systems (HETS) belonging to the 377th Transportation Company, 181st Transportation Battalion, 6th Corps Support Group of the 3rd Corps Support Command also were used to transport the AVLB bridges. Note the Defense Transportation Reporting and Control System (DTRACS) antenna attached on the roof of the cab.

M58A4 MICLIC (Mine Clearing Line Charge). The 3rd Squad has a M113A3 with a M105A2 cargo trailer and a M929 dump truck.

The engineer squads provide personnel and equipment to complete combat engineer missions in support of mobility, countermobility, survivability, and sustaining engineering tasks to increase the combat effectiveness of the supported unit. The primary missions include obstacle construction, obstacle breaching/reduction and engineer reconnaissance. A typical sapper squad with a M113A3 (APC) consists of 8 soldiers: a driver, a tank commander and six dismounted sappers. They are armed with a 7.62mm M240B, a 9mm M9 Beretta, a 5.56mm M249 Squad Automatic Weapon, six 5.56mm M4A1 Carbine, a 40mm M203 Grenade Launcher, and a cal.50 M2 machine gun for the M113A3 APC. They are also equipped with two AN/PAQ-4A Infrared Aiming Lights, two AN/PVS-7B and two AN/PVS-4 night vision goggles, two AN/TVS-5 night vision sights, and one AN/PSN-11 Portable Lightweight GPS Receiver (PLGR).

The assault & obstacle platoon has an Armored Combat Earthmover (ACE) section, a heavy section and a wheeled section that provide special engineer equipment to support mobility and countermobility tasks undertaken by a combat engineer company. The ACE section has six refurbished M9 ACE. Four M60A1 Armored Vehicle Launched Bridge (AVLB), two M200A1 2$\frac{1}{2}$-ton cargo trailers with M58A4 MICLICs, and an M1 Panther II Mine Detection and Clearing Vehicle make up the heavy section.

The "Outlaws" are still equipped with two M48A5 AVLBs and only two M60A1 AVLBs. Depending on the mission, the M60A1 or M48A5 can be used as an armored vehicle launched mine clearing line charge (MICLIC) (AVLM). Mounting up to 2 MICLICs, they are more survivable and maneuverable than the M200A1 trailer with M58A4 MICLIC. The modification of the AVLB into an AVLM is a temporary expedient, and the decision to operate as an AVLM in no way eliminates the responsibility of maintaining a fully operational AVLB.

The wheeled section is composed of two HMMWVs, two M985 HEMTT Cargo truck, two Small Emplacement Excavator (SEE), two M929 dump trucks, and two M796 trailers. In the past, the 54th Engineer Battalion was also equipped with Mini-Flails in the combat engineer platoons, which were returned to the United States for rebuilding. At the moment, only the "Outlaws" and the "Charlie Rock" have a M1 Panther II **(Graphic 3).**

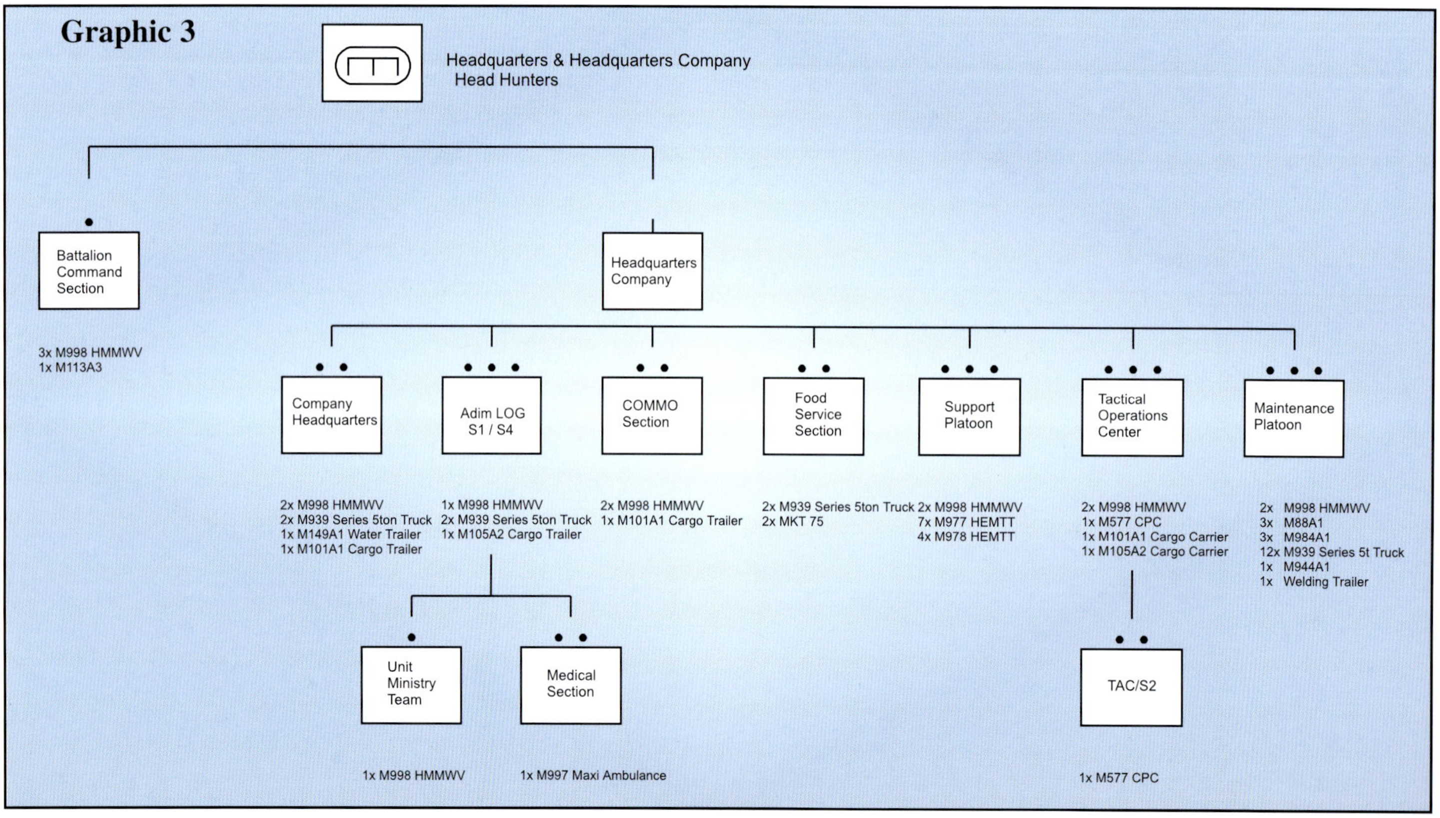

The M60A1 AVLM is used to clear a path through obstacles without having to dismount. The MICLIC (mine clearing line charge) must be prepared for firing before it is moved forward for deployment. This preparation time minimizes the soldiers' exposure to enemy contact. One of the most dangerous tasks an engineer squad can face is dismounting and breaching a minefield while under enemy fire.

The AVLM system consists of a M48A5 or M60A1 chassis, a launcher assembly, M147 firing kit, a M58A4 line charge, and a 12.7cm- (5-inch) MK22 Mod 4 rocket. The 107-meter- (350-foot-) long M58A4 line charge contains 836 kg (1,840 lb) of composition C4. Depending on the mission, one or two MICLIC systems can be installed on the upper hull of the vehicle. Note the nickname "Lazyboy" on the launcher. This M48A5 AVLM was photographed during a recent exercise in the Bamberg local training area. The modification of the AVLB into an AVLM is a temporary expedient and the decision to operate as an AVLM in no way eliminates the responsibility of maintaining a fully operational AVLB.

Primary Engineer Tasks

Mobility

Mobility, countermobility and survivability are the primary tasks of the engineer on the battlefield. Mobility enables the commander to maneuver tactical units into positions of advantage over the enemy. In the attack, engineers reduce enemy obstacles and fortifications that inhibit maneuvering. Expanded engineer reconnaissance capability is needed to identify routes, existing obstacles and minefields. The application of the breaching fundamentals (suppress, obscure, secure, and reduce) and the organization of the force in terms of support, breach and assault force are very important. However, open areas may offer greater opportunity to bypass enemy obstacles because of the greater range of mobility afforded by the terrain.

The engineers must exercise caution when choosing to bypass enemy obstacles since the bypass may lead the force to the enemy's engagement area where they should expect the enemy to make extensive use of rapid mine-laying techniques. When planning a breaching operation, the engineer leaders have to consider bypassing the obstacle as their first option. However, the enemy's intent may be to force the unit to turn or channel into a fire sack. This is why a good reconnaissance of a possible bypass is essential. The unit conducting a bypass must have maximum security during the bypass. If bypass operations are not possible, the options are an in-stride breach, a deliberate breach, an assault breach, or a covert breach.

The breach task force is organized into three elements:

- Support force: Eliminates enemy interference with the breach, isolates the objective area and destroys enemy weapons that are able to fire on the breach force.

- Breach force: Creates lanes in and though the obstacle. While creating the lane through the obstacle, it must also locally secure the breaching site and pass the assault force through the breach.

- Assault force: Destroys or dislodges the enemy force on the far side of the obstacle and seizes the initial foothold in the objective.

To obscure friendly activity and movement, units can use artillery- or mortar-delivered smoke and/or white phosphorous rounds, generators, smoke pots, M203 grenades, chemical units, and burning diesel. After conducting a breach, it is very important to mark the lanes for the following assault force.

Countermobility

Countermobility operations are conducted to augment natural terrain with obstacle systems according to the commander's concept. This adds depth to the battle in space and time by attacking the enemy's ability to maneuver its forces. Engineers advise the commander on the best means to reinforce natural obstacles and to place tactical obstacles that support his plan. Minefields, wire, antitank ditches or berms, road craters, bridge demolition, and all other traditional obstacles may be effective.

Anti-tank ditches or berms require extensive preparation but are very effective when properly prepared. Many areas have irrigation ditches that can be used tactically by the engineers. However, in sandy areas, ditches can be easily filled it. Obstacles are planned to protect the unit's flanks during offensive operations.

The different types of obstacles are:

- Protective Obstacles: They are designed to repel the enemy's assault by placing them around a unit's defensive position.

German Bundesbahn workers direct a M1 Panther II Mine Detection and Clearing Vehicle onto a rail car during the deployment for Operation Iraqi Freedom. The mine clearing rollers (MCRs) and the mine clearing blades (MCBs) were transported via civilian trucks to the harbor at Antwerp, Belgium.

The 1,500-hp Avco-Lycoming AGT-1500 turbine engine provides the 40.2-ton M1 Panther II with a maximum speed of 67km/h (42 mph). But during mine clearing operations the maximum speed should not exceed 16km/h (10 mph). During the conversion of an Improved M1 Abrams MBT into a M1 Panther II, the entire turret (with all hydraulic and electrical systems) is removed to provide space for the different parts of the remote control unit and the communication equipment.

- Tactical Obstacles: They are designed and placed to disrupt, which breaks command and control, frustrates the enemy's timing, interrupts or attacks the enemy when they are massing into formations, and coerces the enemy into the line of the fire.

- Situational Obstacles: These are triggered by an event or situation. Once the general location of the obstacle has been determined, it must be refined to better enhance its effect on the enemy. Situational obstacles should be considered during planning to give the maneuver commander a flexible response to enemy or friendly maneuver.

Survivablility

Survivability is the third primary engineer task. Preparation of fortifications may be difficult and they often require revetments in sandy soils. It may be impossible to dig in rocky areas, arctic regions or locations with a high or periodic water table. Camouflage is very effective when properly employed. However, patterns and techniques must be selected very carefully to blend with the local environment.

Direct-fire weapons are more effective from dug-in positions. Logistic areas require survivability support as these sites are large,

Depending on the mission, the M1 Panther II can be equipped with Mine Clearing Roller (MCR) or a Mine Clearing Blade (MCB). To limit the risk to the two-man crew during mine detection and clearing missions, the M1 Panther II can be operated by remote control with the Omnitech Robotics Standard Robotic System (SRS).

The 3.4-ton Mine Clearing Blade on the M1 Panther II is powered by means of the new Armored Junction Box (AJB), which is located to the left of the driver's hatch. The blade was designed primarily for use on sand, sand-based soil and topsoil. The lifting mechanism of the MCB consists of two identical units. It contains the electric motor used to raise the blade and a speed ear assembly that prevents the blade from lowering too rapidly. Note the chains holding the moldboards, on which loose dirt and mines are gathered for discarding.

Six fuel tanks, with their total capacity of 1,909 liters (504 gallons), provide the M1 Panther II with a cruising range of nearly 430km (267 miles). Although the turbine engine can be operated with nearly all kinds of fuel, the U.S. Army units mostly use JP-8 jet fuel or diesel. The MCB clears a lane about 100cm (3.3 yards) wide in front of the each track, leaving an unploughed lane 150cm (5 yards) wide between the tacks. Through this unploughed area the MCB drags a "dog bone", which is used to detonate tilt-rod mines. In contrast to the MCR, the MCB can be raised for longer marches. When the MCB is in use, the M1 Panther II reduces its maximum speed of 10 km/h (6 mph). If being moved an extended distance, the MCB is loaded onto a transport and then installed.

relatively static, and difficult to camouflage. As a result, they are vulnerable to enemy interdiction. Emphasis should be placed on ammunition and fuel dumps. The engineer's digging assets, such as bulldozers, small emplacement excavators or armored combat earthmovers, should be tasked to provide survivability support to these sites.

With long-range observation and fields of fire, defensive positions are very vulnerable to offensive fire. This fact, coupled with a lack of natural obstacles, may lead the commander to concentrate the bulk of his engineer effort on survivability positions. But coordination between the engineers and the maneuver element is critical to ensure the proper placement of positions. Without effective coordination, much valuable time is wasted in re-digging positions.

In heavy units, initial vehicle fighting-position construction starts with the M9 Armored Combat Earthmover (ACE). But commanders have to consider that the ACE has different digging capabilities than the dozer. In many cases, armored combat earthmovers start with hasty positions, which are then upgraded to improved positions as time allows. Also, the ACEs may start work before corps-level dozers arrive to augment the construction effort. In this situation, the ACEs would dig the hasty position and the dozers would improve them. The platoon must plan for this augmentation to efficiently integrate corps assets into the construction effort.

After having reviewed the organization and the responsibility of the engineers, we will now take a closer look on some of the vehicles used by the 54th Engineer Battalion.

A soldier of the 54th Engineer Battalion armed with a 5.56mm M16A4 with an attached M203 40mm grenade launcher and Aimpoint M68 high performance red dot reflex sight provides cover during an obstacle breach at the CTMC-Hohenfels, Germany.

Each sapper squad has one 5.56mm M249 Squad Automatic Weapon (SAW) gunner. The M249 is an interesting weapon. While its standard ammunition feed is by a 200-round disintegrating belt, it is also capable of firing ammunition from standard M16/M4 magazines inserted in a magazine below the SAW.

Sappers perform a wire obstacle breach during recent training at the CMTC-Hohenfels in Germany. After conducting a breach, it is essential that the breaching force mark the lanes for the following assault force using marking panels.

A sapper prepares a Bangalore torpedo to breach a wire obstacle with mines. (CPT John Hudson)

M9 Armored Combat Earthmover (ACE)

Although the design originated in the late 1950s, the first few prototype Universal Engineer Tractors (UET) were built by the Caterpillar Tractor Company and the International Harvester Company in the early 1970s. When initial testing was completed, the vehicle was type classified as the M9 Armored Combat Earthmover (ACE). Funding of the M9 ACE was slow to materialize and, as a result, the production of the vehicle did not begin until BMY was awarded a construction contract in 1986.

The M9 ACE, which was introduced into the U.S. Army inventory in late 1989, is a highly mobile, armored, combat earthmover capable of supporting forces in both offensive and defensive operations. It is designed to operate in forward areas of the battlefield as a general purpose engineer vehicle. The M9 ACE was designed to provide support in three basic areas – mobility, countermobility and survivability. The vehicles breaches berms, prepares anti-tank ditches, prepares combat roads, removes roadblocks, digs hull defilade fighting positions for guns, tanks and other battlefield systems to increase their survivability, and prepares access routes at water obstacles. The active Army has a total of 447 M9 ACEs.

The engine, drive train and driver's compartment are laid out in the rear of the vehicle, while the front features a 6.7m^3 bowl, apron and dozer blade. Because of its light weight of only 16.33 tons, the M9 ACE is transportable in C-130, C-141, and C5A aircraft. The armor consists of welded aluminium with selected steel and aramid-laminated plates. The vehicle hull is a welded and bolted aluminium structure with a two-speed

As soon as the sappers mark the lane, the assault force is able to move through the wire obstacle. The assault force can then destroy or dislodge the enemy force on the far side of the obstacle and seize an initial foothold in the objective.

Each Combat Engineer Platoon has four M113A3 APCs, a M200A1 2½-ton cargo trailer with M58A3 MICLIC, two M105A2 2 1/2-ton cargo trailers, a M998 HMMWV, a M548 Volcano Mine Dispenser, and a M929 dump truck. Note the ACAV shield around the commander's cupola on this M113A3 armored personnel carrier (APC) and the rucksacks attached to the side of the hull.

Invar winch located at the rear that is capable of 11.4-ton line pull. Towing pintle and airbrake connections are also provided. An armored cupola containing eight 1x vision-blocks covers the driver's compartment and provides good sight in all directions. Just forward of the cupola is a windshield that becomes the front panel of a canvas enclosure that can be erected over the driver's area for use during inclement weather.

The vehicle is equipped with a unique hydropneumatic suspension system that allows the front of the vehicle to be raised, lowered, or tilted to permit dozing, excavating, rough grading, and ditching functions. The principal components are eight high-pressure hydraulic rotary actuators that are connected to the road wheel stations. During high-speed travel, this system assures a smooth ride through the use of shock-absorbing accumulators. In order to use the blade for scraping, the entire front of the hull must be lowered by means of the hydropneumatic suspension, then the blade of the hull's front button edge is used to scrape the ground. To perform this operation, the bowl is usually first filled with 8.2 tons dirt for ballast with the driver selecting clutch brake steering and unsprung suspension settings and then raising the front apron blade. Next he uses the two lever controls to lower the front suspension until the hull scraper digs into the ground just enough for the particular operation. Then he slowly drives forward until the bowl is filled with the desired amount of ballast, after which the apron is lowered and closed. By carefully controlling the height of the suspension, dozing with the blade can be accomplished.

The M9 ACE has armor protection against small arms and artillery fragments, a smoke grenade launcher system, a fire extinguisher system, and a NBC overpressure protection system for the operator. Most of the operators complain about the M9 ACE armament, which consists only of a 9mm M9 Beretta, and would prefer to have a Browning cal.50 M2 machine gun or a 40mm Mk.19 Mod 3 automatic grenade launcher.

The well-known Cummins V903C-T295 8-Cylinder diesel engine with 14.8-liter displacement and 295 hp provides the M9 ACE with a maximum speed of 48 km/h (30 mph) with a cruising range of nearly 470 km (292 miles). The transmission, which offers six forward and two reverse speeds, is a Clark Equipment Company 13.5 HR 3610-2, with torque converter. The steering unit, a Twin Disc, Inc. hydraulic dual gear

The Detroit Diesel 6V53T 6-cylinder engine with turbocharger and 275hp provides the M113A3 with a maximum speed of 65 km/h (40 mph). Note the smoke grenade launcher at the front and the tactical marking of the 54th Engineer Battalion on the side.

A typical sapper squad with a M113A3 consists of 8 soldiers: a driver, a tank commander and six dismounted sappers. They are armed with a 7.62mm M240B, a 9mm M9 Beretta, a 5.56mm M249 Squad Automatic Weapon, six 5.56mm M4A1 Carbine, a 40mm M203 Grenade Launcher, and a cal.50 M2 machine gun for the M113A3 APC.

Each combat engineer platoon has a M113A3 with a M200A1 trailer and M58A4 MICLIC. This vehicle is equipped with the Multiple Integrated Laser Engagement System (MILES) II during an exercise at the CMTC-Hohenfels, Germany. Note the Combat Vehicle Kill Indicator (CVKI) at the front and the marking devices attached to the side of the vehicle.

steer and clutch brake unit, is located under the floor in the alcove at the rear of the ACE and is accessible by lifting the floor plates in that area. Vertical obstacles up to a height of 46 cm (18 inches) and trenches up to a width of 160 cm (63 inches) are no match for the ACE.

There have been a number of System Improvement Plans (SIPs) carried out on the existing vehicles over the past years to improve the performance, durability, readiness, and other characteristics. Some of the upgrades have included improved 200 Amp generators, hardened track pins with modified track shoes, thicker belly plates, and fixes for pulled actuator mounting rings. Steel dozer blades will replace the aluminum blades as the current blades suffer damage when used in rocky terrain and generally wear down more easily. The repair is difficult and involves time-consuming aluminium welding, resulting in extensive mission downtime. The new steel dozer blade is less expensive, more resistant to wear and easier to repair if damaged. The M9 ACEs were also equipped with new road wheels. The center is the same steel wheel as the existing road wheel, but the surface compound is polyurethane instead of rubber. This new wheel is consistently lasting 6-10 times longer than the rubber wheel.

Even though the M9 ACE is effective in supporting the combat troops like armor and infantry on the battlefield, many engineers would like to go back to using their old M728 Combat Engineer Vehicles (CEV), which were removed from the U.S. Engineer inventory during the last few years.

M60A1 Armored Vehicle Launched Bridge (AVLB)

The armored vehicle launched bridge (AVLB), with a length of 11.3 meters (37 feet), a width of 4.0 meters (13 feet) and a height of 3.9 meters (12.8 feet), was designed to launch and retrieve a class 60 scissors-type bridge. It consists of three major sections: the launcher, the main body and the bridge. The launcher is mounted as an integral part of the main body, which is either based on the M48A5 or the M60A1 tank assembly powered

The Mine Clearing Line Charge (MICLIC) system consists of a M200A1 $2\frac{1}{2}$-ton trailer chassis, a launcher assembly, a M147 firing kit, a M58A4 line charge, and a 12.7-cm (5-inch) Mk.22 Mod 4 rocket. The line charge is 106 meters (350 feet) long and contains 2.27 kg (5 pounds) per linear foot of composition C4 explosive.

TA 085

1:35 M1A1 Abrams

Hubert Cance

Each MICLIC trailer transports and fires one charge and then it must be reloaded. The loaded-charge container weighs 1,283 kg (2829 lb), so a lifting device such as a 5-ton wrecker or a HEMTT is needed. An experienced crew can perform the reload in about 20 minutes.

by an air-cooled, turbocharged 750 hp 12-cylinder Teledyne Continental AVDS-1790-2D diesel engine, which provides a maximum speed of 48 km/h (30 mph). Its fuel tanks, which have a capacity of 1,419 liters (375 gallons), provide the M60A1 AVLB with a cruising range of nearly 470 km (292 miles).

The Detroit Allison Diesel CD-850-6A powershift crossdrive transmission has two forward speeds and one reverse speed. With a ground clearance of 35.6 cm (14 inches), the M60A1 AVLB can climb obstacles up to a height of 46 cm (18 inches) and is able to cross trenches up to a width of 214 cm (7 feet). The suspension, which is of the torsion bar type, consists of six road wheels (with the idler at the front and the drive sprocket at the rear), with three track-return rollers. Hydraulic shock absorbers are fitted at the first, second and sixth road wheel stations.

The aluminium alloy girder bridge that folds in half is mounted on top of the main body. By means of a special hydraulic system operated by the driver, the bridge is raised to a vertical position then extend forward, opening as it proceeds forward. Bridge emplacement can be accomplished in 2 to 5 minutes, and retrieval can be accomplished in 10 minutes (under armor protection). Depending on the mission, the bridge can be retrieved from either end. It can also be launched or retrieved either up or down hills with grades of 15% and may be launched on side slopes up to 8% or any combination of up, down and side slopes. The AVLB spans a 15-meter (50-foot) gap for Military Load Class (MLC) 70 and spans an 18-meter (60-foot) gap for MLC 60. The total weight of the M60A1 AVLB without bridge is 41.7 tons and 55.3 tons with bridge.

The crew of the M60A1 AVLB consists of the tank commander and the driver located in the middle of the hull. Two M250 smoke grenade launchers mounted beside the bridge launcher at the front provide the vehicle with self-screening capability.

The 54th Engineer Battalion in Bamberg, Germany, received the first refurbished M60A1 AVLB on 13 November 2002. Under the program 10/20 Inspection/Repair, upgrade and modification of the chassis of the M60A1 AVLB Chassis NSN 5420-00-889-2020 and overhaul of subcomponents when required, 68 old M60A1 AVLBs are being refurbished by the German defense contractor Flensburger Fahrzeugbau GmbH (FFG) to meet today's standards. The improvement program consists of three lots (1st: 23 tanks, 2nd: 20 tanks, 3rd: 25 tanks) and an optional 4th lot of 36 vehicles.

The MICLIC is a rocket-propelled, explosive line charge. It is used to reduce minefields that contain single-impulse, pressure-activated AT mines and mechanically activated AP mines. The MICLIC clears a 14 x 100-meter (46 x 328-foot) path and has a 62-meter standoff distance from the launcher to the detonation point.

Each engineer line company has a M577A1 command post carrier in the company headquarters platoon.

The Volcano Mine Dispenser provides soldiers scatterable mine capability delivered by the M548 tracked vehicle or by helicopters like the UH-60L Black Hawk. One of the two Volcano M548s of the "Outlaws" was damaged and the company was not able to fix the vehicle in time. The second one used a tow bar to tow the first to the railhead in Bamberg, Germany.

The U.S. Army is currently considering a reduction in the number of improved vehicles from 68 to 56 because of funding. The improvements in the areas of bridge, launcher, hydraulic systems, final drives, electrical systems, chemical agent resistant coatings (CARC), and water ration heaters will allow the M60A1 AVLBs to keep up with current Abrams Assault Force. If necessary, the fire extinguishing systems, brakes, fuel systems, and suspension (including the tracks, shock absorbers and the road wheel arms) are also being refurbished. The refurbishment of a single M60A1 AVLB takes nearly 120 to 150 days and is done by FFG in Flensburg, Germany.

The M48A5/M60A1 Armored Vehicle Launched Bridge (AVLB) only supports Abrams tank units making a cautious crossing at reduced gap length and at a reduced crossing speed. The M60A1 Armored Vehicle Launched Bridge entered the U.S. Army inventory in the late 1980s, and the plan is to use the AVLB through 2015 and beyond.

Engineer units employ MICLIC in response to minefield breaching requirements identified by the maneuver unit. A typical mission is as follows: The Engineer Company's MICLIC trailers are moved to a position designated by the maneuver commander and a squad of engineers prepares the line charge. Suitable combat vehicles like the M113A3 tow the MICLIC trailers to the point of the breach. Movement to the firing site and execution is performed under control of the breaching force commander. A second MICLIC may be fired from the breached lane if a longer breach is required. After firing, if additional breaching missions are anticipated, the empty trailer is reloaded with another line charge brought forward from the supply trains. It may also be towed to a pre-selected position and dropped for later recovery by support elements.

The 69.7-ton M104 Wolverine Heavy Assault Bridge System (HABS) was developed to replace the M48A5/M60A1 AVLB in the combat engineer vehicle inventory, but the funding of this program was reallocated to fund the Army Vision transformation strategy after only 46 vehicles had been built. The M104 Wolverine is powered by the well-known 1500 hp Avco-Lycoming AGT-1500 gas turbine.

M60A1 Armored Vehicle Launched Mine Clearing Line Charge (MICLIC) AVLM

The M60A1 AVLM is used to clear a path through obstacles without having to dismount. The MICLIC must be prepared for firing before it is moved forward for deployment. This preparation time will minimize soldiers' exposure to enemy contact. One of the most dangerous tasks an engineer squad can face is dismounting and breaching a minefield under enemy fire. Each lane reduction element is built around an engineer platoon that has been reinforced with M1A1 Abrams main battle tanks with mine rollers or mine ploughs, an M60A1 AVLM and a Bradley Squad.

An engineer platoon may use a pair of M58A4 MICLICs as the primary technique to breach a minefield. The M60A1 AVLM takes up a position 15-20 meters (16-66 yards) behind or to the side of the lead tank and fires one of its two M58A4 MICLICs. As soon as the engineer platoon fires the second MICLIC, the engineer squads employ manual explosive techniques to push additional lanes through the minefield. The lanes are spaced at least 100 meters (109 yards) apart to reduce artillery effects. The firing of the MICLIC will attract enemy artillery fire to the lane location, minimizing risk to the dismounted breach squads. If M1A1 main battle tanks with mine ploughs are available they plough lanes either alongside or in place of the dismounted engineer squads. While the infantry squads provide local security throughout the operation, the main battle tanks move up to detect the forward edge of the minefields and back off 50 meters (55 yards), providing near side security. The engineer squads toward the rear of the formation place far recognition signals to guide follow-on units into lanes. Each lane is individually designated either by color or number to control movement through the breach.

As soon as a MICLIC clears a lane through the minefield, a M1A1 Abrams MBT with Mine Clearing Roller (MCR) is used to proof the lane for unexploded mines.

Commanders and engineers must consider the risk of losing sappers and other breach assets when planning for breaching operations. Even when the MICLIC fires successfully, it can only clear a 100-meter-long path in an obstacle. This is excellent for small obstacles, but not for the deep obstacles found in many breaching operations. The MICLIC has a "skip zone," where mines are left untouched. Also, deeply buried mines, non-pressure-fuse mines, and overpressure-resistant mines are very resilient against the MICLIC.

M1 Panther II Mine Detection and Clearing Vehicle

In June 2002, the 54th Engineer Battalion (C) (M) received the first two M1 Panther II Mine Clearing and Detection Vehicles based on the improved IP M1 Abrams. During Operation Joint Endeavor, the U.S. Army began the development of a remote-controlled mine-detection and -clearing vehicle to counter the mine threat in Bosnia-Herzegovina. At first they used old M60A3 main battle tank hulls for the six prototypes, named M60A1 Panther I Mine Detection and Clearing Vehicle, because of the low costs and the availability. These were equipped with a remote control unit and a mine-clearing roller. After intensive tests it was decided to use IP M1 Abrams MTB hulls for the series production because the old M60A3 hulls were too difficult to maintain and spare parts were in short supply.

A sapper marks the lane through a wire obstacle while a M1A1 Abrams MBT belonging to the assault force moves through to destroy or dislodge the enemy force and seize the initial foothold.

Sappers of the 54th Engineer Battalion employ a wire obstacle during training in the local training area in Bamberg, Germany.

The beret is now the Army's standard headgear in garrison, but the BDU patrol cap is still be worn while participating in field duty or work details, at deployed locations when helmets are not being worn, and on other occasions designated by commanders. Note the blue U.S. Army flash and the crest of the 54th Engineer Battalion, which is worn by enlisted soldiers and NCOs. Officers wear their rank on the flash instead of the crest. The colors of the 54th Engineer Battalion crest are white (argent) and red (gules) for engineers. The three swords (later described as "daggers") symbolize the three outstanding achievements of the unit during World War II: the beachhead invasions of Algeria/French Morocco, Sicily and Normandy. Descent from the 20th Engineer Regiment (2nd Battalion, 20th Engineer Regiment) is indicated by the canton, which bears the coat of arms of that organization and indicates its World War I service.

During the conversion of an IP M1 Abrams into a M1 Panther II, the entire turret (with all hydraulic and electrical systems) is removed to provide space for the different parts of the remote control unit and the communication equipment. To limit the risk for the two-man crew during mine-detection and -clearing missions, the M1 Panther II can be operated remotely using the Omnitech Robotics Standard Robotic System (SRS). Most components of the SRS are mounted on the robotics rack to the left of the commander's seat in the middle of the hull. They provide mobility control for the M1 Panther II by manipulating the vehicle's braking, steering, throttle, transmission, parking brake, and gear-shifting functions under the control of the operator using the Operator Control Unit (OCU). During route-clearing or -proofing operations, the M1 Panther II is the lead vehicle, followed closely by a control vehicle, usually a M113A3 APC or a HMMWV. The operator steers the M1 Panther II remotely with the OCU from the control vehicle by watching video transmitted from the two on-board weatherproof color video cameras and monitoring from the vehicle's voltmeter, tachometer and speedometer.

Depending on the mission, the M1 Panther II can be equipped with a Mine Clearing Blade (MCB) or a Mine Clearing Roller (MCR). During a typical mission, the MCR-equipped M1 Panther II will advance until a mine is detected, either through detonation or sight. Once the mine has been detected, the M1 Panther II backs off and another Panther II, or even a M1A1/M1A2 MBT equipped with a MCB, is brought forward. A path through the minefield is then cleared using the MCB. Once the path is cleared, the MCB will back off, and once again the M1 Panther II with MCR is brought forward to proof the path.

This method of clearing mine fields is ideal. However, there are times when the user will deviate from this method. The MCR was designed to be used primarily on rocky to semi-soft soil, while the 3.4-ton MCB was designed to be used primarily on sand, sand-based soils and topsoils. If you are clearing a loose, sandy area, the MCR may not be very effective. On the other hand, if you are clearing a very hard, rocky, or clay-based soil, the MCB will not be effective. In such cases, the commander will have to rely on either the MCB or MCR as the sole mine-clearing device.

The M1 Panther II powerpack, which includes the turbine engine, the exhaust system and the transmission, provides primary power for tank movement and electrical and hydraulic power for auxiliary systems. The well-known 1500hp Avco-Lycoming AGT-1500 turbine engine provides the 40.2-ton M1 Panther II with a maximum speed of 67 km/h (42 mph). The Detroit Allison Diesel X1100-3B automatic transmission has four forward and two reverse speeds. The advanced rotary shock absorber torsion bar suspension with support rollers and track adjusting links give the M1 Panther II outstanding cross-country mobility and ensures that the tank can follow armor and infantry units from the combined arms teams over every kind of terrain.

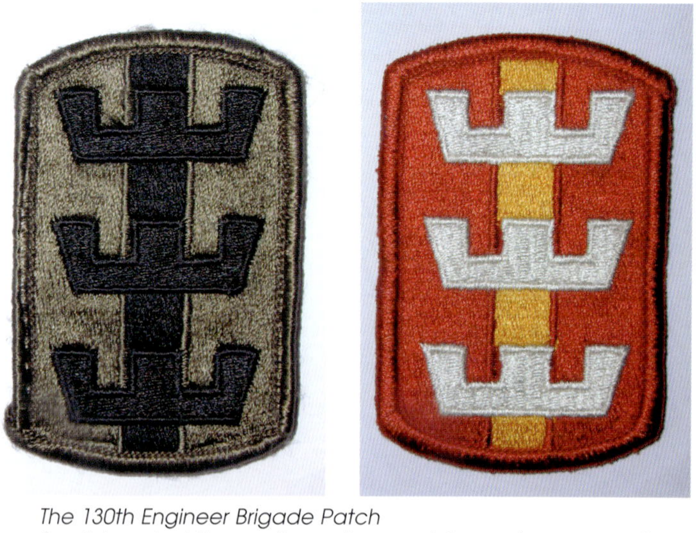

The 130th Engineer Brigade Patch

Scarlet and white are the colors used for engineer operations. The bar refers to a lever, support, fastener, or a measuring device and other facets of engineering operations. The three horizontal tower embattlements, each with three merlons, denote the military aspects of the brigade. In numerology, the number three is the symbol of completeness.

The six fuel tanks, with a total capacity of 1,909 liters (504 gallons), provide the tank a cruising range of nearly 430 km (267 miles). The M1 Panther II armament consists of a Browning cal.50 M2 machine gun with an effective range of 1,200-1,600 meters (1,312-1750 yards) and two M250 smoke grenade launchers, which are located beside the commander's cupola. The vehicle's communication system is made up by two Single Channel Ground and Airborne Radio Systems (SINCGARS) and the AN/VIC-3 Vehicular Intercommunications System for the crew. The hull of the M1 Panther II is protected by an advanced armor similar to the Chobham Armor of the UK Ministry of Defence. The fire extinguishing system installed in the M1 Panther II engine and crew compartment can extinguish any fire within a few seconds.

Despite the current transformation of the U.S. Army, the U.S. Engineers will play a very important role in the future as part of the combined arms team along with the armor, infantry, aviation, and field artillery elements.

The author would like to thank Lieutenant Colonel Ed Jackson; Captain Bryan Sizemore "Headhunters"; Captain John Hudson "Outlaws"; Captain Braden Lemaster "Bulldogs" and Captain Mark Wilson "Charlie Rock" from the 54th Engineer Battalion (C) (M); Mrs. Anne Kelley from the 130th EN BDE PAO; Mr. Gerald Koch from Flensburger Fahrzeugbau GmbH; Captain Steve Nettleton and Master Sergeant John Cavanaugh from the Viper Combat Camera Team (HHC CMTC); Mr. Nick Damario from the 279th BSB PAO; Lieutenant Colonel Brian McNerney from the USAREUR PAO; Mrs. Millie Waters from the IMA-Europe PAO; and especially all "Dagger Battalion" soldiers for their outstanding continuing support. He also wishes them good luck and all the best during their current deployment in Operation Iraqi Freedom in support of the global war on terrorism.

"Safety First." The soldiers of the 54th Engineer Battalion have to wear safety vests during rail loading operations.

The Body Armor Set Individual Countermine (BASIC) is designed to provide protection to the soldier conducting countermine clearing operations by reducing the effects of blast and protecting against fragments from small to medium-size anti-personnel (AP) mines and booby traps.

The 54th Engineer Battalion Crest

Colors are white (argent) and red (gules) for engineers. The three swords (later described as "daggers") symbolize the three outstanding achievements of the organization in World War II: the beachhead invasions of Algeria/French Morocco, Sicily and Normandy. Descent from the 20th Engineer Regiment (2nd Battalion, 20th Engineer Regiment) is indicated by the canton, which bears the coat of arms of that organization and indicates its World War II service.

An engineer of the 54th Engineer Battalion in Bamberg, Germany, uses an AN/PSS 12 Mine Detecting Set to detect mines in the remaining lane between the M1 Panther II tracks. The white plastic strips are used to mark the mines on the ground. Head protection is provided by a combined installation of the standard PASGT helmet with a soft ballistic cover. To provide optimal visual clarity, the face shield is constructed with a clear polycarbonate lens that provides reduced protection against fragments.

M1 Panther II Mine Detection and Clearing Vehicle

Ralph Zwilling

The MCB clears a lane about 100cm (39.4in) wide in front of each track, leaving an unplowed lane 150cm (59in) wide between the tracks. Through this uncleared area, the MCB drags a "dog bone", which is used to detonate tilt rod mines.

Mines have become prominent weapons in warfare because they are simple devices, very effective, inexpensive, readily manufactured anywhere, easy to lay, and very difficult and dangerous to find and destroy. The production costs of an anti-personnel mine are between US$3 and US$30. The current cost of clearing only one mine is US$300-US$1000. The United Nations estimates that more than 100 million mines pose significant hazards in more than 68 countries around the world. Currently, about 2-5 million new mines are laid every year. Stockpiles exceeding 100 million mines are held in over 100 nations, and 50 of these nations produce additional 5 million new mines every year. The rate of clearance is far slower. It is estimated that 800 people are killed and 1,200 maimed each month by land mines. The primary victims are unarmed civilians, mostly children. To counter the mine threat faced by troops and civilians in an area of responsibility (AOR), modern engineers need systems to clear mine-affected areas efficiently, economically, rapidly, and reliably.

Development

The mission of the U.S. Forces in Bosnia-Herzegovina revealed that the available mine-detection and mine-clearing systems in the inventory were not effective at removing the mines in a reliable and quick manner. There were _ to 1 million mines along the zone of separation in this country. If minefields in the Balkans were cleared only by non-mechanical methods, it would take decades instead of years to remove this threat to life and limb. During the Operation Joint Endeavor, the U.S. Army began the development of a remote-controlled mine-detection/mine-clearing vehicle to counter this mine threat in Bosnia-Herzegovina.

Because of low costs and availability, they used the hulls from old M60A3 main battle tank (MBT) for the first six prototypes, which were equipped with a remote-control unit and a mine-clearing roller. The M60A3 turret was completely removed to make enough space for the components of the remote-control unit. These first prototypes were named M60A3 Panther I Mine Detection and Clearing Vehicle. The M60A3 Panther I worked very well with the M728 Combat Engineer Vehicle (CEV) in the Balkans. The operator of the remote control had adequate height and a good line of sight to monitor the Panther's progress during proofing operations. The CEVs blade was also used to improve the effectiveness of the proof by skimming the ground when it followed the Panther. It also provided an excellent recovery vehicle for the Panther when is got stuck. Finally, the CEVs boom was used to download the mine roller system from the transport vehicle, which eliminates the need to have a crane on site to prepare the Panther for operation.

Today, the M728 CEV is no longer in the U.S. Army inventory and the M88A1 is relied upon to lift the roller. After intensive tests, the Army decided to use old IP M1 Abrams hulls for the series production in the future since the old M60A3 hulls are too difficult to maintain and spare parts are in short supply. The first two M1 Panther II Mine Detection and Mine Clearing Vehicles were delivered in June 2002 to the 54th Engineer Battalion "Dagger Battalion" of the 130th Engineer Brigade based in the Warner Barracks, Bamberg, Germany.

Technical Information

During the conversion of an Improved M1 Abrams MBT into a M1 Panther II, the complete turret, including all hydraulic and electrical systems, is removed to provide space for the different parts of the remote control unit and the communication equipment. The driver's station at the front center of the vehicle remains virtually unchanged. It is equipped with a monitor showing the condition of vehicle fluid levels, batteries and electrical equipment. The Mine Clearing and Control Panel (MCCP), which is used for controlling the mine-clearing equipment on the front of the M1 Panther II, is installed to the left of the driver's seat.

During Operation Joint Endeavor, the U.S. Army began the development of a remote-controlled mine-detection and mine-clearing vehicle based on the M60A3 hull to counter the mine threat in Bosnia-Herzegovina. After intensive tests, they decided to use old IP M1 Abrams hulls for the series production in the future since the old M60A3 hulls are too difficult to maintain and spare parts are short in supply.

The Mine Clearing Roller (MCR) consists of two roller banks with two push arm assemblies that operate independently of the other, freely following over the contours of the terrain it encounters. Each roller bank has four rollers, which apply ground pressure higher than that exerted by the tank. This principle ensures the explosion of pressure-fused anti-tank mines, which would otherwise explode under the track itself.

There are three observation periscopes or two periscopes on either side, as well as a central image intensifying periscope for night vision in the driver's hatch. This Driver's Vision Enhancer, the AN/VSS-2(V)2A, enables the tank crew to maneuver at normal daytime driving speeds in darkness and in poor visibility conditions such as dust, smoke and other battlefield obscurants. The commander's station is situated in the middle of the hull directly behind the driver. The commander's seat is mounted on the new floor plates that cover the torsion bars of the Abrams suspension.

In contrast to the modern M1A1 and M1A2 Abrams MBT, the M1 Panther II Mine Detection and Clearing Vehicle is not equipped with a Nuclear, Biological, Chemical (NBC) over-pressure ventilation system to protect the crew in the chemical warfare field. Because of this, the crew has to wear Mission Oriented Protective Posture (MOPP) IV suits and M42A1 chemical-biological masks in case of an NBC threat or attack, which can cause problems in a hot and humid climate environment.

To limit the risk for the two-man crew during mine-detection and mine-clearing missions, the M1 Panther II can be operated by remote control using the Omnitech Robotics Standard Robotic System (SRS). The SRS equipment mounted aboard the tank consists of the Vehicle Control Unit (VCU), four High Integration Actuators (HIA), a Video Transmitter Unit (VTU), two color video cameras (CAM) with Manual Pan/Tilt Units (MPTU), a Microphone (MIC) Box, two System Input/Output (SIO) units,

The 1500hp Avco-Lycoming AGT-1500 turbine engine provides the 40.2-ton M1 Panther II with a maximum speed of 67km/h (42mph), but during mine-clearing operations, the maximum speed should not be higher than 16km/h (10mph).

The 9-ton MCR was intended primarily for detecting rather than clearing minefields on rocky to semi-soft soil. The MCR should not be used to clear a loose, sandy area. A secondary use of mine rollers is to proof lanes created by other means.

This photo shows to advantage the details of the rear of the M1 Panther II. Note the tactical bumper code of the 54th Engineer Battalion, which belongs to the 130th Engineer Brigade, and the T-158 track with replaceable track pads.

a Safety Radio Unit (SRU), a Manual/Auto Switch Box, a Speedometer/Tachometer Unit, a Transmission/Throttle Interface Unit (TIU), an Uninterruptible Power Supply (UPS), an antenna mast with three antennas, and the M1 control interfacing mechanisms.

Most components of the SRS are mounted on the robotics rack to the left of the commander's seat. This provides mobility control for the M1 Panther II by manipulating the vehicle's braking, steering, throttle, transmission, parking brake, and gear-shifting functions under the control of the operator using the Operator Control Unit (OCU). Countermine functions controlled by the SRS include Power On, Blade/Roller select, Roller release, Right/Left Blade select and Raise/Lower Blade. The operator also commands the starting and stopping of the vehicle's engine.

The SRS provides several safety features that enhance the operations of the remote-controlled vehicle. It stops the M1 Panther II automatically if radio communications between the Operator Control Unit (OCU) and the on-board systems are lost or disrupted or if any of the safety-critical equipment on board the vehicle malfunction. A red panic button is provided on the OCU for the operator to initiate a stop with a single, simple action in case of loss of control or suspected problems. The SRS also comes with an independent safety radio system (SRS) that can be used to provide an emergency-stop capability that overrides all SRS commands. The operator can override the safety radio system itself before the start of remote operations, if desired.

The Manual/Auto Switch Box mounts to the left rear of the driver's seat, a location that allows access through the driver's hatch, which is

The skidshoe adjusting plate allows for three plowing depths: 20, 25 and 30cm. (7.8, 9.8 and 11.8in). However, for peacetime operations the MCB should not be used at a depth greater than 20cm. As the tank pushes the MCB through a minefield, loose dirt and mines are rolled up onto moldboards and to the sides to clear the area in front of the tracks.

It takes nearly 30-45 minutes to install the Mine Clearing Blade (MCB) on the hull of the M1 Panther II. Note the M250 smoke dischargers and the antenna mast. This tank belongs to Assault & Obstacle Platoon of the Alpha Company "Outlaws", 54th Engineer Battalion.

hinged to open upward and swung starboard. When the switch is set to AUTO, the M1 Panther II can be operated remotely. If it is set to MANUAL, the M1 Panther II can only be operated from inside the tank. During route-clearance or proofing operations, the M1 Panther II is the lead vehicle. It is followed closely by a control vehicle, usually a M113A3 APC, or a High Mobility Multipurpose Wheeled Vehicle. The control vehicle (CV) contains the Panther II operator, the Operator Control Unit (OCU) and the Safety Radio Transmitter (SRT). A microphone that can be positioned within the vehicle enables the operator to hear through a speaker in the OCU the engine sounds and the sound of the actuators when the engine is not running.

The operator steers the M1 Panther II remotely with the OCU by watching video transmitted from the two on-board, weatherproof, color video cameras and by monitoring feedback from the vehicle's voltmeter, tachometer and speedometer. The front CAM pan angle is intended to provide the vehicle driver's view, while the rear CAM pan angle is intended to look straight behind the vehicle. Both video and feedback information are displayed on the OCU simultaneously, in real time, to give the same view and information as if the operator was actually sitting in the vehicle's driver's seat. The compact keyboard on the OCU is for contractor maintenance and repair purposes only.

Mine Detection and Mine Clearing

Depending on the mission and the soil, the M1 Panther II can be equipped with the Mine Clearing Blade (MCB) or the Mine Clearing Roller (MCR). As the M1 Panther II was introduced to the army inventory not long ago, the 54th Engineer Battalion in Bamberg is still testing to determine how to use the tank most effectively. Current plans dictate that the M1 Panther II should primarily be used to the rear of the forward line of own troops (FLOT) to secure and clear supply routes, assembly areas, and airfields from anti-personnel and anti-tank mines. U.S. Army officials also think the vehicle could be used in task force missions with armor and infantry battalions. In this case, a MCR-equipped M1 Panther II would clear the roads in front of the combat force with a speed of 16km/h (10mph). At this speed, the 9-ton MCR is capable of detonating or neutralizing single-pulse, pressure-activated anti-tank and anti-personnel mines laid on the surface or buried up to 16cm (6.3in) deep in the track path of the pushing tank.

The MCR is comprised of two roller banks with two push arm assemblies that operate independently of each other, freely following over the contours of the terrain it encounters. Each roller bank has four rollers, which apply ground pressure higher than that exerted by the tank. This principle ensures the explosion of pressure-fused anti-tank mines, which

The lifting mechanism of the MCB consists of two identical units: the electric motor used to raise the blade and a speed ear assembly that prevents the blade from lowering too rapidly. Note the chains holding the moldboards.

The six fuel tanks, with their total capacity of 1,909 liters (504 gallons), provide the tank with a cruising range of nearly 430km (267 miles). Though the turbine engine can be operated with nearly all kinds of fuel, the U.S. Army units mostly use JP-8 jet fuel or diesel.

The operator steers the M1 Panther II remotely with the OCU by watching video transmitted from the two on-board, weatherproof, color video cameras and by monitoring feedback from the vehicle's voltmeter, tachometer and speedometer. Note the red panic "stop" botton on the OCU.

would otherwise explode under the track itself. Each roller bank rolls a width of approximately 110cm (43in), and there is an unrolled area of nearly 150cm (59in) between the rollers. The Anti-Magnetic Mine Activating Device (AMMAD) system, which is connected between the roller banks, detonates magnetic-fused anti-tank mines, which are laid in the full width of the vehicle. The same device will also activate tilt-rod-fused mines, which are laid between the track paths.

The mounting kit for the MCR includes a hydraulic system, two quick-disconnect mechanisms, two outboard mounting brackets, a center-mounting bracket, two tie rods, and linkage between the mounting brackets. Power to the MCR is supplied from the mine roller harness protective cover, which is located just behind the right front light. The MCR-equipped M1 Panther II will advance until a mine is detected, either through detonation or sight. Once the mine has been detected, the M1 Panther II backs off and another Panther II or even a M1A1/M1A2 MBT equipped with a MCB is brought forward. A path through the mine field is then cleared using the MCB. Once the path is cleared, the MCB will back off and, once again, the M1 Panther II with MCR is brought forward to proof the path.

This method of clearing mine fields is ideal. However, there are times when the user will deviate from this method. The MCR were designed for use primarily on rocky to semi-soft soil, while the 3.4-ton MCB was designed to be used primarily on sand, sand-based soils and topsoils. If you are clearing a loose, sandy area, the MCR may not be very effective. On the other hand, if you are clearing a very hard, rocky, or clay-based soil, the MCB will not be effective. In such a case, the commander will have to rely on either the MCB or MCR as the sole mine-clearing device. It should be noted that the Mine Clearing Blade is limited in a depth of 30.5cm (12in). The skidshoe adjusting plate allows for three plowing depths: 20, 25 and 30cm (7.8, 9.8 and 11.8in). However, the MCB should not be used at a depth greater than 20cm for peacetime operations.

During route-clearance or proofing operations, the M1 Panther II is the lead vehicle. It is followed closely by a control vehicle, usually a M113A3 APC or a High Mobility Multipurpose Wheeled Vehicle.

This "over-the-shoulder" view provides a look at the image on the screen of the OCU that enables the operator to remotely control the M1 Panther II.

The M1 Panther II armament consists only of a Browning cal.50 M2 machine gun with an effective range of 1,200-1,600 meters (1,312-1,750 yards) and two M250 smoke grenade launchers that are located besides the commander's cupola.

The Mine Clearing Blade is powered through the new Armored Junction Box (AJB), which is located right of the driver's hatch. As the tank pushes the MCB through a mine field, loose dirt and mines are rolled up onto moldboards. As the tank continues to push the MCB, dirt and mines are pushed to the sides, clear of the tracks. As in case of the MCR, a chain and AMMAD assembly, attached between the moldboards, detonates any tilt-rod mines in the center lane. The control box in the driver's compartment provides operator control for raising and lowering of moldboards. A manual-release system allows the moldboards to be lowered to the plowing position in the event of electrical system failure. Unlike the MCB, the MCR cannot be raised during long road marches but there is a quick-disconnect to release the rollers in case of an attack or an emergency. During long marches, the rollers have to be transported by a truck such as the M985 HEMTT. It takes between 30-45 minutes to install the roller assembly with a M88A1 Armored Recovery Vehicle. As cargo space inside the M1 Panther II is very limited, a special stowage basket was attached behind the commander's hatch. In series, the M1 Panther II is equipped with the T-158 track with removable track pads, which the M1A1 and M1A2 main battle tanks also use.

It is noteworthy that the Panther II is painted at the production line with only a forest green Chemical Agent Resistant Coating (CARC) and not in the NATO three-color-camouflage paint scheme. Two red marking sticks attached to the rear of the tank are used to operate the tank in reverse mode with the camera and the OCU.

Specifications

The M1 Panther II powerpack, which includes the turbine engine, the exhaust system and the transmission, provides primary power for tank movement and electrical and hydraulic power for auxiliary systems. The 1500hp Avco-Lycoming AGT-1500 turbine engine provides the 40.2-ton M1 Panther II with a maximum speed of 67km/h (42mph). An engine change can be performed in an hour, as compared with four hours for the diesel engine in an M60A3 MBT. Time between overhauls (TBO) is estimated to be 1800 hours, about three times greater than that for a diesel engine. The modular construction of the turbine also makes routine maintenance much easier and faster. Engine components can be changed without having to dismantle the entire engine, thus fewer parts are required to be on hand.

The Detroit Diesel Allison X1100-3B automatic transmission has four forward and two reverse gears. The advanced rotary shock-absorber torsion bar suspension with support rollers and track adjusting links gives the M1 Panther II outstanding cross-country mobility and ensures that the

The M1 Panther II crew consists of the driver, who is situated at the center front of the vehicle and the commander, who sits directly behind him. The yellow markings of the 54th Engineer Battalion symbolize a bridge.

This photograph shows to advantage the cal.50 M2 machine gun and two M250 smoke dischargers that make up the total armament of the M1 Panther II. The two tow ropes are secured around the commander's cupola, which is originally from the M113. Note the color video camera below the barrel of the M2 and the stowage boxes for the smoke grenades.

The mounting kit for the Mine Clearing Roller (MCR) in detail. It includes a hydraulic system, two quick-disconnect mechanisms, two outboard mounting brackets, a center mounting bracket, two tie rods, and linkage between the mounting brackets. In front are the quick-disconnect mechanism and the protective cover for the mine roller harness connector.

tank can follow armor and infantry units in every terrain. The six fuel tanks, with a total capacity of 1,909 liter (504 gallons) provide the tank a cruising range of nearly 430km (267 miles). Though the turbine engine can be operated with nearly all kinds of fuel, the U.S. Army units mostly use JP-8 jet fuel or diesel. Because of the high ground clearance of 47cm (18.5in), vertical obstacles up to a height of 124cm (4ft) and trenches up to a width of 274cm (9ft) are no match for the M1 Panther II, but only when no mine-clearing system is installed. Additionally, water obstacles up to a depth of 122cm (4ft) can be crossed without preparation.

The commander's cupola, which has five vision blocks, is originally from the well-known M113 APC. The M1 Panther II's armament consists of a Browning cal.50 M2 machine gun, which has an effective range of 1,200-1,600 meters (1,312-1,749 yards), and two M250 smoke grenade launchers, which are located beside the commander's cupola. The vehicle's communication system is composed of two Single Channel Ground and Airborne Radio Systems (SINCGARS) and the AN/VIC-3 Vehicular Intercommunications System for the crew.

The crewmembers wear the Combat Vehicle Crewman (CVC) helmets, which represent a convergence of technologies without compromise to offer high ballistic performance, durability, comfort and improved communications. The Kevlar ballistic shell provides V50 ballistic test performance at a minimum of 1,400 fps (427 Mps), while typical performance is 1,500 fps (456 Mps). M1 Panther II crews use this helmet series with the M-138/G microphone and MK-1697/G headset unitized for a comfortable adjustable liner. A three-position switch allows radio, intercom and listen-in-only modes. Sound attenuation is maximized via ear cup design, and outbound communications benefit from a microphone signal-to-noise ration of 15 dB. Padding materials provide comfortable protection from temperature extremes while significantly attenuating external noise. The hull of the M1 Panther II is protected by an advanced armor similar to the Chobham Armor by the U.K. Ministry of Defence. The fire-extinguishing system installed in the M1 Panther II engine and crew compartment can extinguish any fire within a few seconds.

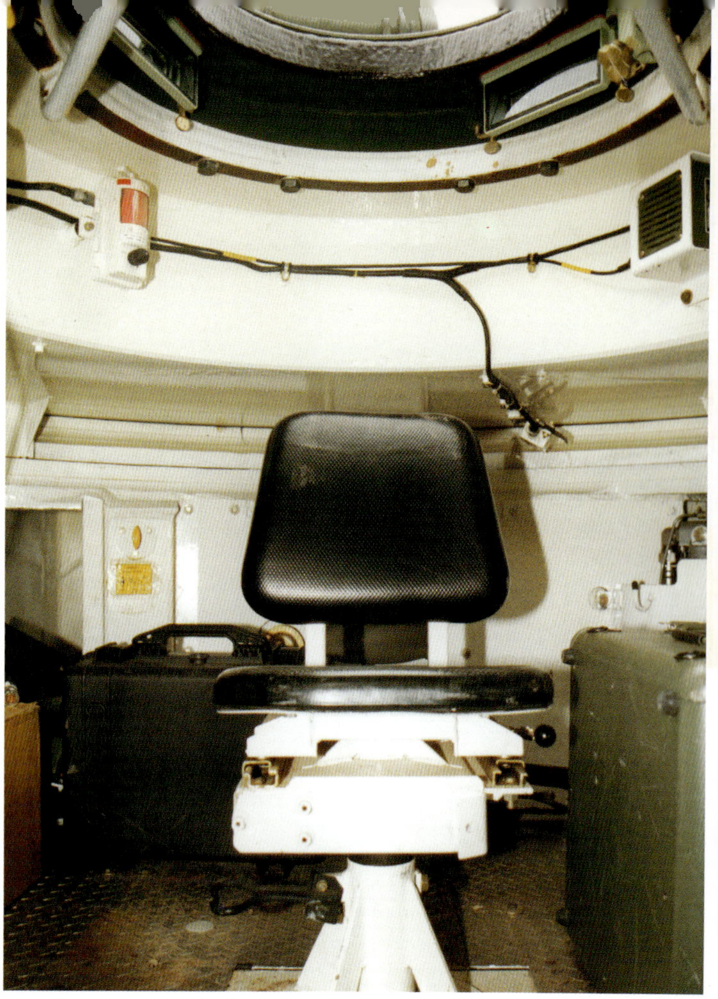

The commander's station in the middle of the hull. Note the dome light, the speaker of the SINGCARS radio system and the fire sensor of the fire extinguishing system located behind the seat.

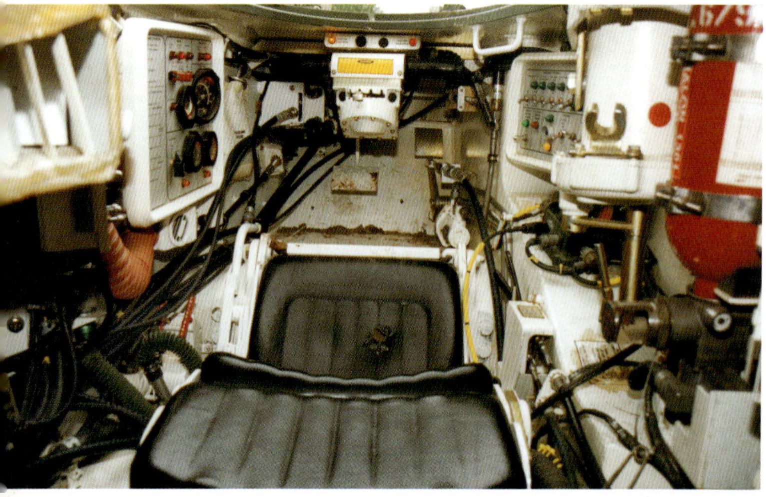

A close-up of the driver's station. The main driving control is a motorcycle-type-steering T-bar that can be positioned in a number of ways depending on the wishes of the driver. The seat is lightly padded with a black ribbed leather-like covering, while the rest of the interior is painted the same gloss white as the fighting compartment. The main control panels are beside the driver's seat.

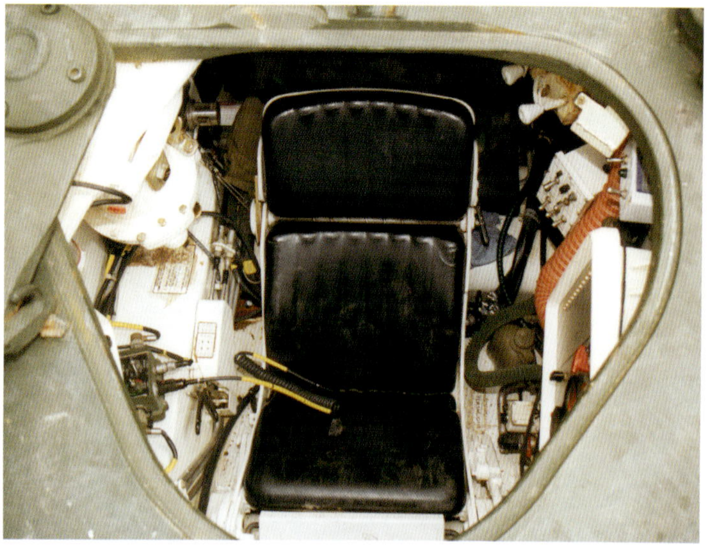

The driver of the M1 Panther II sits in a reclining position in the front center of the hull surrounded by his controls. The seat is very comfortable, providing excellent back support, and can be adjusted in a variety of directions.

This photo shows some components of the Standard Remote System (SRS) situated beside the commander's station. The equipment is pre-assembled at the factory to allow for quick and easy installation in the field.

The Single Channel Ground and Airborne Radio System (SINCGARS) is designed on a modular basis to achieve maximum commonality among the various ground and airborne system configurations. A common Receiver Transmitter (RT) is used in the manpack and all vehicular configurations. The system operates on any of the 2320 channels between 30-88 megahertz and is designed to survive in a nuclear environment.

With the issuance of the M1 Panther II Mine Detecting and Clearing Vehicle, the U.S Engineers received the first tank that was especially built to detect and clear mines. At present, the U.S. Army has a need for another 12 M1 Panther II for their engineer units. A corps engineer company equipped with the new M1 Panther II is divided into a company headquarters section, two combat engineer line platoons, and an assault & obstacle platoon. The company headquarters section, which is equipped with two HMMWVs, a M577A1 Command Post Carrier, a M113A3 armored personnel carrier (APC), and a M939 series 5-ton truck, provides command, control and communication to the company.

Both of the combat engineer line platoons have a platoon headquarters with a HMMWV and a M113A3, and three engineer squads. The first squad is equipped with a M113A3, a M548 Volcano mine dispenser and a M105A2 1_-ton cargo trailer. The second squad has a M113A3 towing a M200 Trailer with a M58 MICLIC (Mine Clearing Line Charge) and a Mini Flail. A M113A3 with a M105A2 1_-ton cargo trailer and a M929 dump truck make up the third squad. The assault & obstacle platoon has an ACE section with six M9 Armored Combat Earthmover (ACE), a heavy section with four M60 AVLB, two M200 Trailer with M58 MICLIC, and a M1 Panther II. Depending on the mission, the MICLIC can be mounted on the M60 AVLB. The wheeled section is equipped with two HMMWVs, two M985 HEMTT, two SEE (Small Emplacement Excavator), and two M929 dump trucks.

The author would like to acknowledge the outstanding support provided by CPT Hudson (CDR A Company, 54th EN BN), the soldiers of A Company "Outlaws", 54th Engineer Battalion, Mrs. Kelley (130th EN BDE PAO), 279th BSB PAO, and the U.S. Army Europe Public Affairs Office.

Because the cargo space inside the M1 Panther II is very limited, a special stowage basket is attached behind the commander's hatch. Note the mount for the SRS antenna on the left and the SINCGARS radio antenna on the right.

The mounting frame supports all parts of the system and is attached to the front of the M1 Panther II. The Mine Clearing Blade (MCB), which is powered through the Armored Junction Box (AJB), is controlled by the Mine Clearing and Control Panel (MCCP). It is installed to the left of the driver's seat.

Urban Camouflage of the French Armor

Yves Debay

Brightly standing out against the green of the trees in the background, the new camouflage scheme definitely does not lend itself to stealth in the countryside.

More than one third of Western Europe is made up of urban areas. This means that potential conflicts in the 21st century will take place within cities. The conventional point of view regards a city as a location that can reduce the potential of a superior enemy force. Along with this view one must also consider the negative ways urban areas may be used in the conflicts of tomorrow: suburban sanctuaries that shelter the movement of terrorists, zones of illegality that are controlled by organized crime, and ethnic ghettos that escape the control of the State. Just as the military leaders during the Cold War prepared to counter an attack of 50,000 tanks by the Warsaw Pact nations on the plains of northern Europe, the French *Armée de Terre* (ground forces) felt the need to study the potential threats that exist in the new century.

A newly camouflaged AMX-10 RC reconnaissance vehicle belonging to the 4th Squadron of the 1st Spahis Regiment is seen on a training sortie.

Within the framework of fighting in urban environment, the STAT (*Services Techniques de l'Armée de Terre* – Technical Services of the Land Forces) experimented with a new type of camouflage for their armored vehicles. The prestigious 1st Spahis Regiment was charged with field testing this new type of camouflage, which is similar to what the British applied to their Chieftain tanks of the Brigade of Berlin in the 1970s. Just returned from Afghanistan, the 4th Squadron will be covering its twelve AMX-10 RCs with the new livery.

The new camouflage consists of large squares of two shades of gray, one of ochre and one sand-colored tan shade of paint. Since the paint is water soluble, running the tank through a shower will reveal the traditional three-tone central European paint scheme or the two-tone desert camouflage. If the experiment proves satisfactory, every armored vehicle of the *Armée de Terre* engaged in urban fighting might adopt the new camouflage scheme.

This frontal view of an AMX-10 RC shows the gray paint on the top surface of the vehicle and the tan, ochre, light gray and dark gray colors applied to the front.

The 1st Spahis AMX-10 RC travels on an asphalt road as it exits a wooded environment on its approach to a city.

This photograph shows how the new camouflage looks in an urban setting.

With its 105mm gun and its new urban camouflage, this AMX-10 RC is a formidable weapon for use in urban conflicts.

This side view of a 1st Spahis AMX-10 RC shows to advantage the several-colored squares that compose the new French urban camouflage paint scheme.

This VBL (Véhicule Blindé Léger – *Light Armored Vehicle), seen in the foreground, belongs to the 2nd Squadron of the 1st Spahis Regiment. Note the predominance of the ochre and tan colors applied to the side of the vehicle.*

The VBL and AMX-10 RC of the 1st Spahis resume their training sortie. It is apparent that the gray color painted on the flat surfaces of the vehicles would camouflage well with the gray asphalt of the road when viewed from the air.

The 1st Spahis is a famous regiment of the 6th BLB (Brigade Légère Blindée – *Light Armored Brigade), which maintains the tradition of the cavalry of the North African Army. The 1st Spahis was involved in the Gulf War, Bosnia and Kosovo.*

The way that this VBL almost blends with the building standing behind it illustrates the wisdom in adopting the new urban camouflage paint scheme.